# *Guide to* WELLNESS *Through* STRETCHING

### Change your range & improve mobility. Get ready to change your life!

## BY DALE DEIS & ED STILES

◆ FriesenPress

One Printers Way
Altona, MB R0G 0B0
Canada

www.friesenpress.com

ISBN
978-1-03-911258-2 (Hardcover)
978-1-03-911257-5 (Paperback)
978-1-03-911259-9 (eBook)

*1. MEDICAL, ALLIED HEALTH SERVICES, PHYSICAL THERAPY*

Distributed to the trade by The Ingram Book Company

# Table of Contents

# Foreword

Years ago, I was approached by a keenly intense Physical Therapist who said that an article I had written about posture made him think we should work together. He handed me some thoughts he had chicken scratched on a piece of paper. At the time, in the mid-nineties, they were revolutionary—approaching the body as one unit rather than compartmentalized areas to train or rehab in isolation, therapy that ignores alignment will fail, correcting osteoarthritis through movement, individualizing therapy rather than using an assembly line approach... the list went on.

After an hour's "interview," several things were very apparent:

Dale Deis had an unceasing drive to critically analyze the inner workings of the human body.

He was unwilling to accept the status quo, instead forever seeking a better approach to therapy and training.

He was absolutely passionate about healing people. I knew that if we combined our talents, we could help even more of them.

So, our journey began—sharing an office, resources, successes, and failures/lessons. This book is the culmination of his thirty years as a Physical Therapist and my own thirty years as an Exercise Physiologist.

This is not just another book about stretching, but a guide to help readers feel, move, and function much better. We wish you all the best on your wellness journey and we are honoured to be a part of it.

Ed Stiles
BPE, Clinical Exercise Physiologist

# Introduction

Few interventions in the fitness and rehabilitation realm can elicit dramatic effects like that of a simple stretch when applied properly. "Thank you, that stretch changed my life," "I haven't tied my shoes," "I haven't gotten out of bed without back pain," "I haven't walked without foot pain," "I haven't played my sport pain free in years." We have heard countless similar testimonials shared by clients over a combined sixty years of experience, and we need to share our insight toward helping people move, feel, and function better through stretching.

This book lays a foundation of knowledge so that you can understand why stretching is the key to health. It explains how the body adapts to its consistent application, which over time will improve mobility bringing the vast benefits of stretching to light. We will dispel myths, misunderstandings, and explain common mistakes to avoid. We will share the subtle "secret" that allows muscles to actually lengthen and adapt in a static stretch. Then we apply it all through easy to understand photos and descriptions so readers can embrace it themselves and experience the transformative benefits that stretching provides. Read on if you are willing to invest a little time and energy—stretching will truly change your life.

# Stretching

How many times have you been told to "just go and stretch it out"? You scratch your head and proceed to do something that feels "stretchy," and then off you go to your chosen endeavour. Stretching is much more than simply pulling your fingers back or bouncing up and down while trying to touch your toes.

Increasing the flexibility of the short and tight tissues (muscles, fascia, nerves, and blood vessels) through stretching is recognized as the key to restoring and improving the health and balance of the human body. Doctors, physical therapists, chiropractors, massage therapists, and kinesiologists all encourage stretching to improve the mechanical alignment of the human body. This alignment is ultimately what we seek.

Being able to move freely with no restrictions can feel like an awakening if function has been lost for any extended period of time. Over the course of an active life, injuries will happen, with potential adaptations affecting our ability to move efficiently. Inappropriate footwear, muscle imbalances and sustained postures can also influence our ability to move freely. But injuries can actually afford a learning opportunity. By stretching and regaining normal movement, injuries can be "dealt with," and you will be able to avoid unnecessary disability and pain down the road.

## The Secret

Sad but true, most people stretch incorrectly. For years the fitness industry has been inundated with the overriding concept of "no pain, no gain." Folks who invest time stretching often push their tissues past the stretch reflex and they feel tension, but really, we need to convince them that tension can be the muscle contracting, not relaxing. This is, in fact, the "secret" to static stretching; muscles must be relaxed in order to actually lengthen.

The stretch reflex is a protective mechanism that fires an emergency contraction the millisecond we go too quickly or too far into a stretch. This is why you shouldn't bounce when static stretching. A dynamic stretch gets the muscle ready to be used and activated while a static stretch deactivates the muscle and allows it to relax and lengthen. To realize the life altering changes that stretching provides, this relaxation "secret" must be embraced.

## Types of Flexibility Training

There are two basic types of flexibility training: Dynamic stretching and Static stretching.

Dynamic stretching: The dynamic stretch could be considered a warmup and is done prior to activity. Full arcs of movement are encouraged and muscles, ligaments, tendons, and capsules are stimulated and neurological patterns are rehearsed. Blood flow is encouraged to get ready to perform an activity, whether in the realm of athletics or getting ready to do some gardening. The body is preparing to move and muscles are activated and turned on. The dynamic stretch is an evaluation on the operation of the human body and utilizes the range of motion currently available.

Static stretching: The static stretch is done primarily after activity when the body is nice and warm and more amenable to lengthening of both contractile (muscles and blood vessels) and non-contractile tissues (ligaments, tendons, joint capsules, and fascia), trying to increase available range of motion (ROM). The static stretch can increase the available range of motion at a joint, and due to the principle of joint angle specificity, the body then has to learn how to manage and control this new movement pattern. This is why static stretching before an event is not the best idea.

Turning down the neurological input, which could lead to muscle spasms and protective muscle activation, is yet another goal of static stretching. Static stretching also helps push metabolic waste out of the muscles. The feeling after stretching can be like a weight taken off your shoulders, and can almost feel as if you're floating.

# Benefits of Stretching

We've seen stretching improve so many lives in so many ways. Benefits include improving posture and flexibility, preventing injuries, realigning fascial disruption due to injury or habitual posture, improving nerve mobility and joint nutrition, enhancing muscle blood flow, and reducing stiffness in muscles, tendons, and ligaments as we age.

## Improves Posture

To a very large degree, our health is influenced by how we carry ourselves against the force of gravity. Great posture requires very little effort to maintain, the body is in balance. The postures we assume when we sleep, sit, stand, and move cause the tissues to adapt. It is this adaptation that yields many muscular and tissue changes that can pull joints out of their ideal alignment and "best fit" positions. Some muscles chronically shorten while others lengthen. Adaptations like bone spurs, joint degeneration, and osteoarthritis can result from poor alignment.

Any sustained position under tension will create decreased blood flow (hypoxia) into the muscles. A head-forward posture, for example, will cause decreased blood flow into the back of the neck and upper back, which can cause pain. The lack of blood flow also promotes anaerobic metabolism of these muscles, which then creates lactic acid, causing that familiar burning along with the pain of hypoxia. Bringing the head back up to neutral allows blood flow to return to the muscles, and taking the head and neck back into extension will stretch the structures in the front of the neck that have tightened, resulting in better balance, improved blood flow, and no pain.

Stretching is the first step towards achieving better posture, bringing joints back to balance, and preventing and correcting many postural faults. Improved mobility will allow the muscles to perform their function more efficiently.

# Improves Flexibility

Staying flexible allows us to bend and twist with control to function and perform daily activities easily, but also to be able to absorb impact and sudden movements when necessary, such as bumps and falls or getting hit on the field of play.

The benefits of fluid, pain-free movement are numerous. Getting up pain-free and moving from position to position effortlessly is a great feeling. A bit of dedication and consistent effort is all you need to maintain flexibility. Gravity is always trying to push us back into the ground and any gains achieved can be lost the next day, but stretching consistently allows us to teach the muscles and joints how to move in the best way possible.

# Prevents Injury

The ability of every joint in the body to withstand and recover from injury is directly related to its ability to travel through a full, natural range of motion (ROM).

It just makes sense that if the joints can't travel through a full ROM in a controlled setting, then we are definitely more prone to injury when we ask these joints to go to extremes when moving at the speed of life. Bend and absorb, but don't break by bringing the joints into better balance and improved mobility.

# Realigns Fascial Disruption Due to Injury

Normal muscle, tendons, ligaments, and joint capsule fibres are arranged in parallel fashion. With injury, scar tissue is haphazardly laid down to mend the area. As the scar heals, it will shrink and contract and eventually restrict movement if it does not return to its normal ability to lengthen and shorten. Stretching will help reorient these fibres back to parallel, promote improved blood flow, and return the muscle to suppleness and free movement required for normal function.

# Improves Nerve Mobility

With injury and disuse, nerves can become entrapped in scar tissue or in the small tunnels they run through. Neural information from the brain to the target tissue and back is aided by the flow of fluid in the nerves. Scars and adhesions can inhibit this flow. Stretching can help to release scars and adhesions and help to realign muscle fibres and fascial tissue to allow movement of all relevant structures, including the nerves.

# Improves Joint Nutrition

Most synovial joints in the body (joints that have lubrication), have what is known as a "screw home mechanism". Movement creates stress on the joint surfaces to encourage improved ability of cartilage to absorb forces. Pressure pushes fluid out, then through the middle range pressure is reduced allowing fluid to enter the cartilage. This process creates a pumping of nutrition in and waste out of the cartilage. With uneven force or decreased ROM, this is lost. Stretching buys it back.

## Enhances Muscle Blood Flow

Attaining full ROM through stretching also benefits the muscle because with movement of the joints, there is also movement of the muscles and capsules supporting the joint. Muscles are nourished by arteries, and waste (by-products of metabolism) is then taken away by veins and lymphatics. When these tubes are stretched, the inside of the tube (lumen) gets smaller as the vessel is lengthened and gets bigger when released, creating a pumping action. The muscle is also surrounded and contained by fascia and as the muscle is lengthened it, too, pushes blood out, eliminating waste products, and when released, will allow fresh blood to perfuse the area, bringing in oxygen and nutrition. A good demonstration of this fact is to observe the pinkness of your palm with the hand relaxed. Now, with the palm facing up, bend the straight fingers back with the other hand. Note the blanching of the palm and immediate flush upon release. (By holding a stretch 60 seconds or more, the lack of blood flow (the blanching effect), can cause hypoxia in the muscle tissues which can cause pain, inhibiting relaxation). **Stick to 30-60 second holds, or four to six long breaths.**

## Maintains Muscle Belly Length

As we age, the collagen fibres in the tendons and ligaments get stiffer. Stretching allows us to maintain the length of the muscle belly, putting less pressure on the ligaments and tendons, and causing fewer injuries to these structures. Ligaments and tendons have blood flow for nourishment. Constant pressure inhibits this blood flow and makes ligaments and tendons brittle and easy to injure. Stretching will keep things supple.

# What Happens When We Static Stretch Properly?

*At the muscular level:*
A static stretch is a sliding elongation of the overlapping protein fibres deep in the tissue past their current resting length. If this elongation is applied consistently enough the muscle unit responsible for movement, the sarcomere (the actual contractile unit of a muscle cell), elongates as well. During static stretching, the muscles are consciously down-regulated or relaxed, and this can allow for increased sarcomere length.

After static stretching is performed, the muscles will be able to do things that they haven't been able to do for a long time. Keeping the head up and balanced, for instance, will be a new function for the neck muscles and can cause these muscles to get stiff and sore. It may be difficult to get things moving at first and even when we stretch properly we may experience some soreness. When joint position and range of motion is changed the muscles have to begin functioning in an unfamiliar manner, your "new normal."

Feelings of soreness and stiffness will eventually go away and is part of the rehabilitation process. It is important to realize that you may be undoing years of effective ways of moving through an injury that may not have been efficient, but seemed less painful at the time. Over weeks, months or even years, these same strategies are not as effective as they once were and may lead to pain. Changing habits is difficult and learning new movements is challenging. Stretching is about diligent consistency and then maintenance into the future.

*At the fascial level:*
Fascia is the complex network of tissue that interweaves individual muscle fibres, entire muscles, and their neighbors. Indeed, the entire body is internally bound together with this amazing connective tissue. It connects your baby toe to your pinky finger, which is why it has been hypothesized that flat feet can cause headaches. If

the muscles are under conscious control when stretching, it seems that a significant portion of the stretch effect is, in fact, on the fascia.

Fascia has a strong tendency to shorten and tighten due to age, the cold, poor posture, and muscular imbalance. Normal fascia has very high water content and is "fluid," allowing muscles to slide upon themselves and surrounding tissues. Stretching drives the ability to move better, and movement is lubrication.

# Why Don't We Stretch?

It is very unfortunate that so much misinformation surrounds the role of stretching in health. The popular media has not helped as they sell more papers when stories challenge the status quo. Stories in the 90s questioning stretching made international news headlines by sensationalizing flawed studies and leaving out crucial information that actually supported stretching. Unfortunately, the damage has been done and people have been stretching less ever since.

In our 60+ plus years of training literally thousands of clients, very few of them stated a goal of increasing flexibility. People's primary goals revolve around how they look, not how they function. "Lose the potgutsky, muffin tops, and bubble butt, tone the triceps, inner thighs, and pop the pipes."

Stretching, which yields little to no visible aesthetic change to our bodies, is rarely prioritized in a world where we want quick-fix magic solutions with minimal time and effort.

Let's face it: in the wrong hands, stretching actually hurts; picture a crotchety "old school" coach barking at you, "I said touch your toes!"

People hear conflicting information regarding how to stretch. Some have merit and work better in certain situations, some not so much. People are unsure how to do it properly. This leads to confusion and the impression that you're maybe better not doing it than doing it wrong or wasting your time.

Indeed, the very reason for writing this book is because the people who take the time to stretch are often missing the mark. While they are generally doing some good, they are far from maximizing the time they invest because they often break the rules of optimal stretching.

Also consider that most animals stretch. How many times have you noticed your dog or cat stretch after a nap? Yoga has postures that are examples of nature; up dog, down dog, cat pose and cobra amongst others. Animals tend not to do things that are either a waste of time or energy, as they have fleeting amounts of both, yet most stretch. It must be important.

## How to Stretch

### The Five Rules of Effective Static Stretching

1. Stretching should not hurt. No gain if pain.

2. The muscle must be relaxed in order to lengthen. Take the muscle to the point of mild tension. The less pressure, the better. Do not go too far or too fast.

3. Position your body such that the line of pull of the stretch is functional, and maintain neutral spine when indicated. Stay balanced, controlled, and stable so as to take pressure, tension, and gravity off the target tissue.

4. Take the muscle to the point of comfortable tension and hold. If within 30-60 seconds, the stretch feeling dissipates, push the stretch a bit further. Do the stretches 2X consecutively with a short break between stretches. Each stretch should not exceed 60 seconds. After the first attempt, there will be an increase of blood flow which can enable a better stretch on the second attempt. There is a point where the joint/muscle will not allow any increase in motion, and then it is a simple matter of maintenance.

5. If the stretch feeling does not go away, back off, and try a lower range of motion.

Note: A general cardio warm-up or specific body part warm-up is recommended prior to any stretching, but is not absolutely necessary. Static stretching can still be done at any time.

# Rules of Effective Dynamic Stretching

All of the static stretches can be used dynamically as a mobilizer by USING the muscle being stretched and going rhythmically into and out of the stretch.

Utilize light to progressively more vigorous movement of the muscle for 5-10 repetitions.

1. Do not force the stretch, but encourage functional movement through full range of motion.

2. Move from positions of healthy posture.

3. Always remember: **no gain if pain**.

# The Stretches
## The Core

The core is defined as all of the spinal vertebrae including the lower back, middle back, the neck, the pelvis, and the ribs, as well as the muscles that help to control their movement, not forgetting the blood vessels, nerves, and fascia. The arms, legs and head are connected to the core. The human spine is an amazing mechanical system with 24 vertebrae stacked precariously atop one another, supported by an intricate network of muscles to keep things in place... and on top of this is perched a 5-7 kilogram ball. Things go wrong whenever there is imbalance in the system. Use the following assessments to identify imbalances and use these same movements as stretches to correct them.

## Assessments for The Lumbar Spine

These assessments are used to determine the integrity of the facet joints, (joints at the back of the spine). Can your lumbar spine handle extension and compression? Are the facet joints inflamed? Can your spine handle leaning to either side? Is the range of motion the same to the left as it is to the right?

Note: If you have pain during these assessments, you should seek further medical support.

### 1. Standing Extension and Side Flexion (Oblique Abdominals)

Equipment needed: Bench, table or cupboard

It's important to keep the feet together so movement is in the spine and not the hips.

*Difficult*
Try to touch the back, inside of the knee on the same side unassisted as part of the assessment. This is a dynamic

stretch of the oblique abdominal muscles in the front on the opposite side and a test for the facet joints in the back of the spine on the same side.

If you have no pain doing this maneuver, hold for 30-60 seconds, 2X each side, as a static stretch with your hand on a bench, or repeat 10X as a dynamic mobilizer without the hand on the bench. The ultimate goal is to touch the crease on the inside, back of the knee. If you look over the shoulder at the same time, you are affecting the entire spine from the neck down to the lower back. Healthy joints will be able to tolerate this compression.

### Intermediate

Standing extension with assistance: If there is any pain while leaning backward, put a hand on a bench, table, or even the back of your leg to control the pressure. The goal is to do this unassisted.

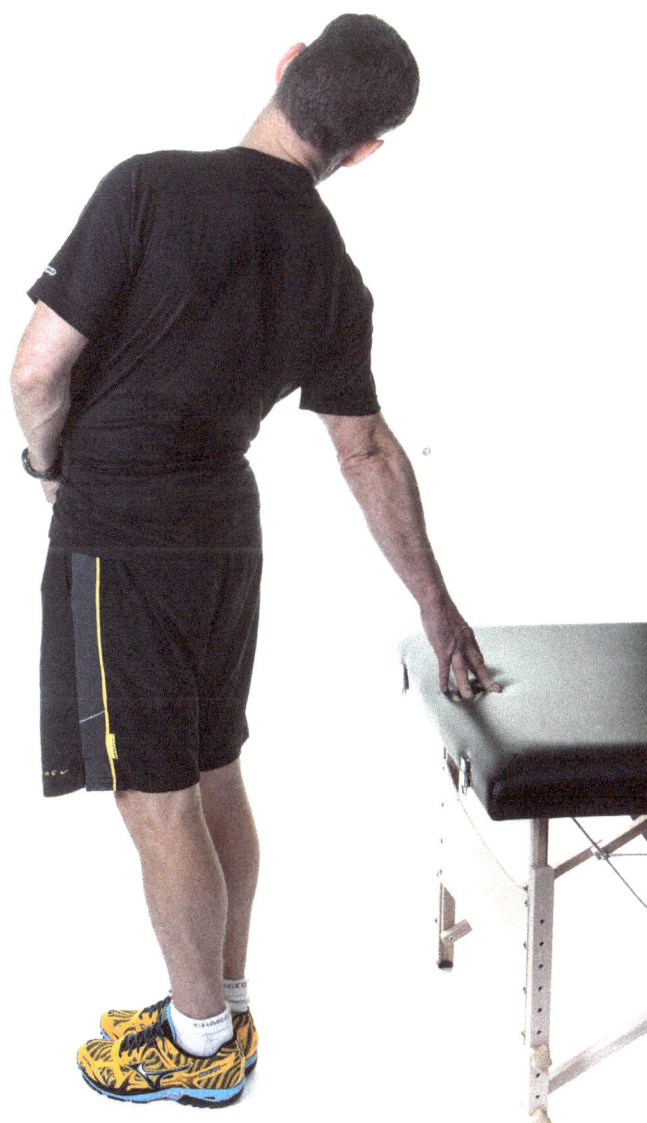

**CORE**

## 2. Side Flexors Stretch (Quadratus Lumborum/Muscles on Either Side of Spine, Including Rib and Neck Muscles)

Equipment needed: Floor and a bench or cupboard

### Ideal
The ultimate goal is the ability to touch the head of the fibula (the bony bump on the outside of the lower leg just below the knee). If you can't do it, use the assisted version below. Once you can reach the head of the fibula, the stretch must be repeated and maintained. Hold for 30-60 seconds, 2X as a static stretch with your hand on a bench, or repeat 10X as a dynamic mobilizer.

### Assisted (if Ideal is not attainable)
Side flexion with assistance: Place a hand on a bench to allow the muscles on the opposite side to relax and stretch. Bend the elbow to control the amount of stretch.

# Spine Rotation

Equipment needed: Exercise mat, sturdy bench or bed

## 3. Lying Torso Rotation (Lumbar)

The lumbar spine is better at side flexion than rotation so try to lengthen the side of the torso rather than trying to simply rotate.

*Beginner*

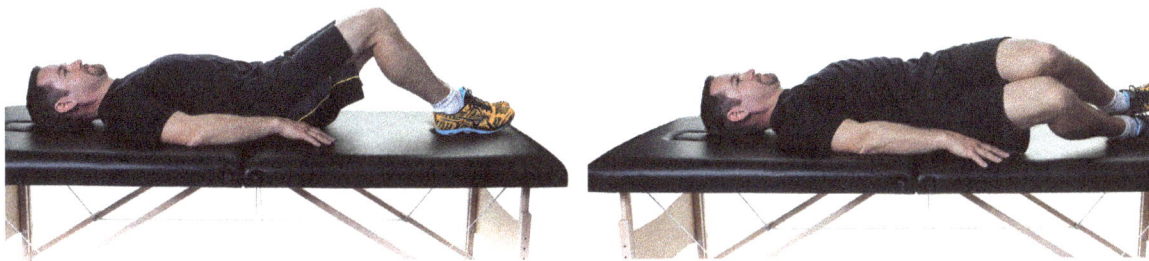

1. Lay on the back with both knees bent to about 45 degrees. Stretch arms straight down at the sides.

2. Slowly rotate both legs to one side keeping the knees and feet together and the opposite shoulder blade in contact with the floor.

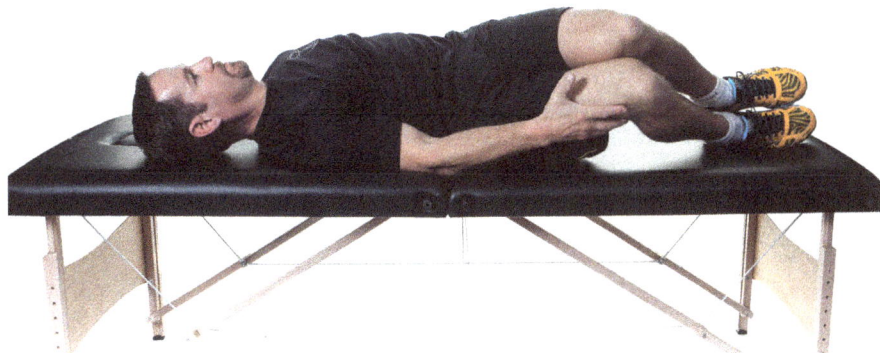

3. Once you can touch the outer thigh to the floor, proceed to the Advanced level. Support with a hand under the lower thigh enables the muscles on the top to relax. Elongate through the top side of the torso, opening the distance between the lower ribs and the top rim of the pelvis. Hold for 30-60 seconds 2X, or repeat 10X as a dynamic mobilizer.

## Advanced

Lying on the floor, start with both knees bent at 45 degrees. Cross one foot over the opposite knee and lower this knee toward the floor. Maintain contact with the opposite shoulder on the floor. Use the weight of the upper leg to bring the knee down toward the surface as far as comfortable. Try to elongate through the side of the lower torso—you are not trying to encourage purely rotation, but rather a combination of side flexion with minimal rotation.

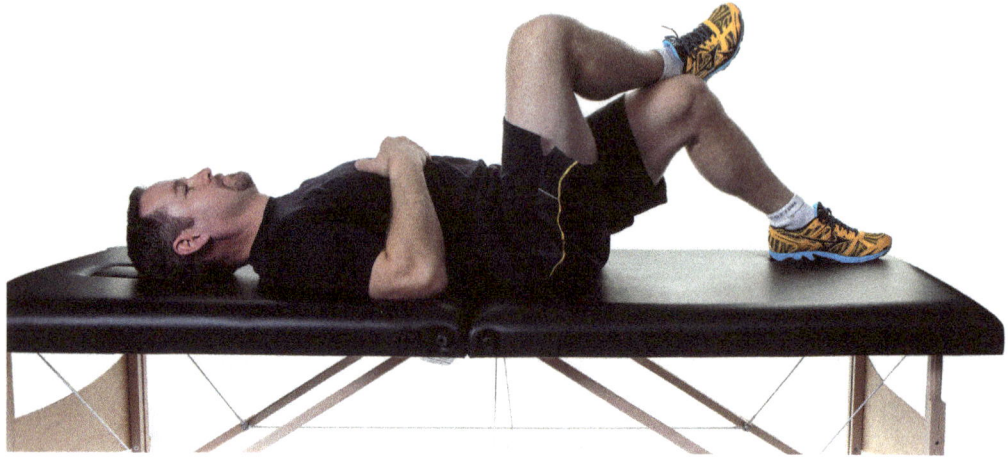

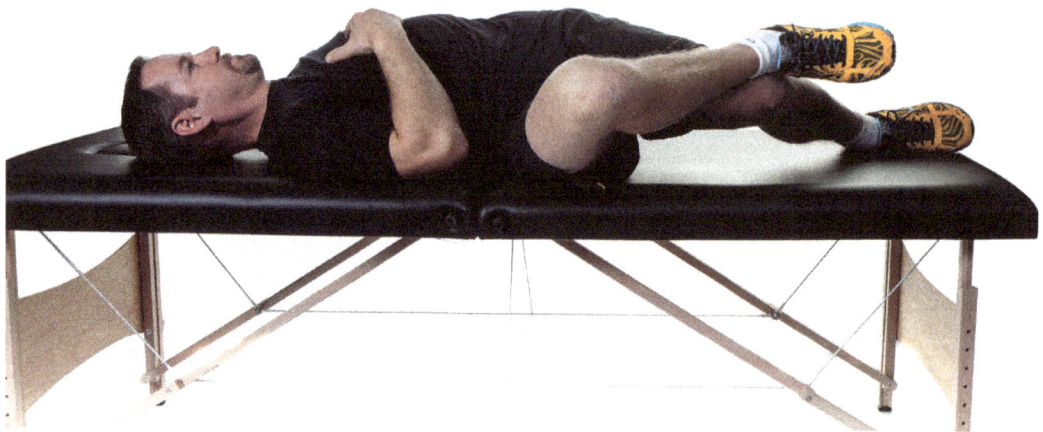

# Lower Back (Lumbar Spine)

## Low Back Extensor Muscles

These are the muscles in the back of the vertebrae, specifically the lower back. They are used to stabilize the vertebrae in lifting, and are also used to come up to standing after bending forward or helping to control the pelvis when getting up from a full, flat footed squat. The following is a progression of stretches for the lower back muscles from the easiest to the most difficult.

### 4. Lying Pelvic Tilt (Beginner)

Equipment needed: Exercise mat, carpeted floor, or bed

The pelvis is the centre of gravity of the entire body. Control of the pelvic girdle allows control of the vertebrae above and the powerful legs below, and is the link between the two. Tightness in the hip flexor or spinal muscles can contribute to an increase in the curve of the lower back.

The ability to control the amount of tilt of the pelvis has huge implications for the health of the lumbar spine.

Lie on your back with knees bent. Inhale and slowly arch your back off the floor to accentuate an anterior pelvic tilt. This is not the position that the human spine needs to spend a lot of time in, but it is important for people to learn the one extreme so the opposite is easier to feel. You may feel jamming in the back. These are facet joints (joints of the spine) being compressed in the back of the spine and if aggravated, they will hurt when compressed too aggressively. Do not go to pain.

Exhale as you contract the buttocks so as to pull your tailbone down toward the heels, effectively tucking the buttocks under, creating a posterior pelvic tilt. At the same time, try to contract the abdominals while trying not to use the legs. A cough will help to engage the abdominals. It should feel as though you are pulling your pubic bone up toward your ribs with the lower back pressing into the floor. This will stretch the muscles of the lower back and when the back has problems, this can feel extremely tight.

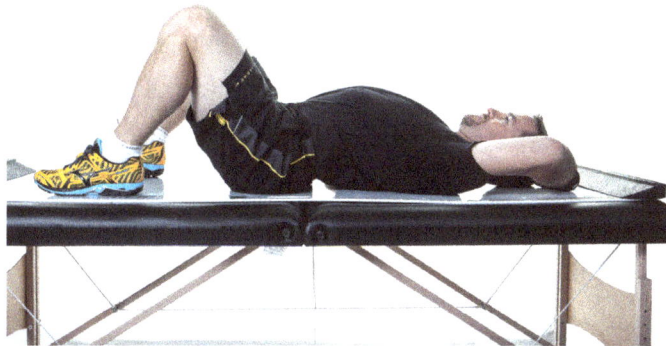

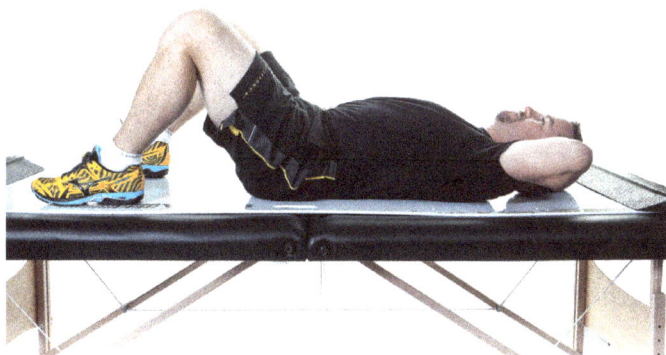

Do not lift the buttocks off the floor. Keep the chin pulled down to try to flatten the neck. Do not push too hard, no pain. Hold for 30-60 seconds as a static stretch, 2X, while continuing to breathe, or hold 2 seconds and repeat 10 times as a dynamic mobilizer.

## 5. Standing Pelvic Tilt (Intermediate)

Equipment needed: Floor

Control of the pelvis in lying is important, but control in standing is even more important if you want to walk or stand efficiently. Inability to maintain a pelvic tilt when standing upright may indicate tight hip flexors. You may want to start with your back against a wall to enhance feedback as this allows you to feel the lower back press into the wall.

Stand with feet shoulder-width apart and pointing straight forward. Bend the knees slightly, lift the chest, pulling the chin back into the throat, maintaining neutral spine. Inhale, and then slowly exhale as you contract the abdominals and buttock muscles to flatten the spine while pulling the pelvis up in the front. Don't let the knees move. Maintain the contraction and keep breathing. Hold for 30-60 seconds, 2X with a pause between, or repeat 10 times, holding 2 seconds and releasing each time, as a dynamic mobilizer. As you inhale and slowly release the contraction of the buttock muscles and abdominals when doing the dynamic mobilizer, allow the lower back to arch a bit.

# 6. Knees to Chest (Beginner)

Equipment needed: Exercise mat, carpeted floor, or bed

The muscles that run along the back of the body are, as a group, responsible for extending the spine and rotating the trunk. They are prone to tightening in people that tip their pelvis forward (think high heels or pregnancy/postpartum) resulting in a big arch in the lower back.

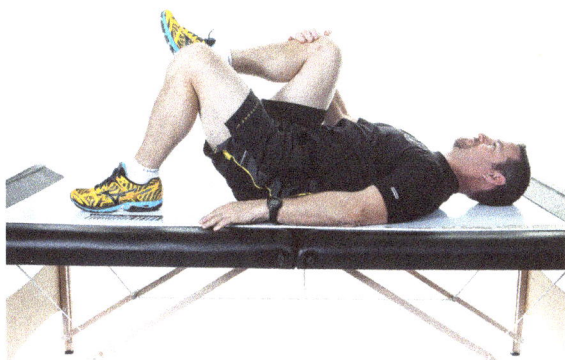

1. Start on your back with knees bent, lift one knee to 90 degrees and grab it with one hand.

2. Then lift the other to 90 degrees and grab it with the other hand.

3. With both hands on the knees, inhale, and then exhale as you gently pull both knees together toward the chest, raising the buttocks off the floor.

4. Keep the head down and the chin pulled in, keeping the neck flat. Do not lift the head.

Note: Move one knee at a time when you're lifting the legs up and down. Try to assist pulling the knees into the chest by using the lower abdominals in the manner of doing a reverse crunch or pelvic tilt. Keep the knees close together. Hold for 30-60 seconds, 2X as a static stretch with a short break between, or do 10 repetitions as a dynamic mobilizer holding 2 seconds each time. Inhale as you let the knees move away from the chest then exhale as you pull the knees back into the chest.

CORE

# 7. Child's Pose (Beginner)

Equipment needed: Floor or exercise mat

Child's pose is another great stretch for the lower back extensors.

1. Kneel on a mat, sit onto heels as far as comfortable with toes pointing as pictured.

2. Walk hands forward as far from shoulders as comfortable.

3. Relax head onto floor

4. Keep the knees together as much as possible to maximize the stretch, and support the trunk weight with the arms. Some may find it easier to keep knees apart to do this stretch. Hold for 30-60 seconds.

**CORE**

# 8. Reverse Abdominals - Hands Down (Intermediate)

Equipment needed: Exercise mat or carpeted floor

This is an amazing lower abdominal strengthening exercise but is also a fantastic lower back stretch and can be used to determine if some specific lower back stretches may be necessary.

1. Start on your back with knees bent, lift one knee then the other to 90 degrees. Hands down on the floor beside you.

2. Exhale as you bring the knees up toward the chest.

## 9. Reverse abdominals - hands on shoulders (Difficult)

Equipment needed: Floor or exercise mat

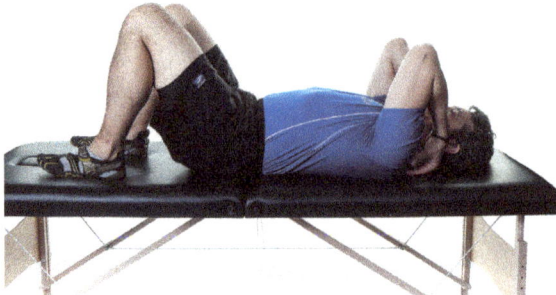

1. Start on your back with knees bent

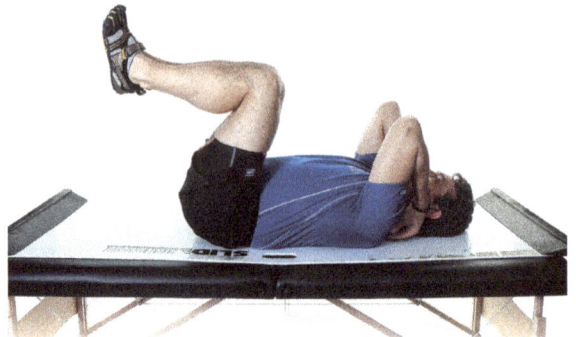

2. lift one knee then the other to 90 degrees

3. Lift the feet higher to straighten the knees a bit to make it easier.

4. Bring knees up to elbows. By keeping heels on the buttocks, this exercise can be made much more difficult.

Start with 5-10 repetitions, as these are surprisingly difficult. You may get tightness in the lower back as the buttocks come off the floor and the back rounds. Try to touch the knees to the elbows without lifting the head or shoulders.

CORE

# 10. Squats (Assisted, Intermediate)

Equipment needed: Leg of squat rack, doorknobs of an open door, edge of the sink, or anything relatively stable.

Squats are an excellent stretch and strengthening exercise for the lower back and they are a fundamental strengthening exercise as well as an amazing stretch of the buttocks, quads, and calf muscles.

1. Stand with feet together and hold something for support. Keep heels down. With feet apart, the spine will remain neutral.

2. Start by pushing the knees forward over the toes and follow with the hips, knees, and ankles bending, coming down until the buttocks are on your heels or as far as comfortable, progressing deeper as your body allows. Use caution if you have hip, knee or ankle pain and try to minimize the pressure. Control the pressure by holding door knobs or the edge of a sink or even a pole to pull up on. If you're knees are really bad, put a chair on either side of you and put the hands on the seat of the chairs to provide support and control the compression. Near the bottom of the squat, you will feel the buttocks dip under and the lower back will curve. This is where the lumbar stretch happens. Eventually you will be able to do this squat with no assistance. Keep the head over the ankles and stay tall. Hold as a static stretch for 30-60 seconds, 2X.

3. If the back is particularly tight, go down with the heels lifted and start on your toes to take the pressure off the lower back. By pumping the heels up to release the lower back tension, then down until tension is felt, the lower back muscles can loosen up with repetition. The squat will provide a nice lower back stretch. Keep the head forward.

Repeat the squat 10-20 times as a dynamic mobilizer. This will strengthen the lower back and the buttocks.

CORE

# 11. Squats (Unassisted, Difficult)

Equipment needed: Wall or door

This is the same as the assisted squats, but this time reach forward with the arms and put the finger tips on a wall. You must have reasonably healthy knees, hips, and ankles to do the full, flat-footed squat and as mentioned above, assistance will help with progression.

Without bending the fingers, wrists, elbows or moving the shoulder blades forward, try to squat all the way down while keeping the heels down.

This will take a ton of practice, but a full squat is a great indicator of overall flexibility from the head to the ankles.

# Trunk Flexors (Abdominal Muscles)

Equipment needed: Exercise mat, carpeted floor, or bed

## Abdominal Stretch

The trunk flexors are prone to shortening and tightening when we hold ourselves in a forward slouch position. These are the muscles used to stabilize the spine when lifting or running, as well as the prime movers to get us up from lying down or returning to upright after bending backward.

## 12. Prone Extension

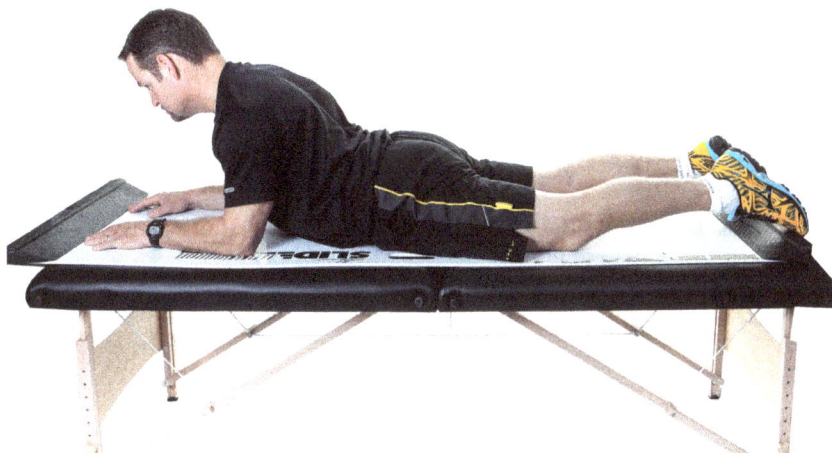

1. Lay face down on a mat to establish the starting position. Place elbows directly beneath shoulders with hands straight out in front and you're looking like the Egyptian Sphinx.

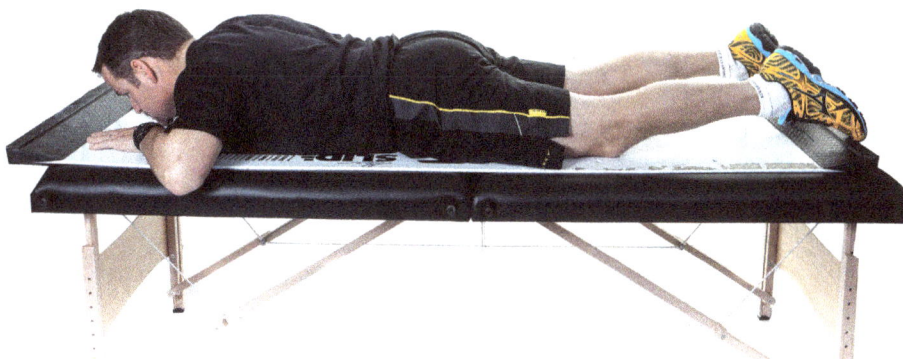

2. Allow the elbows to go out to the side, and lower chest to floor without moving the hands.

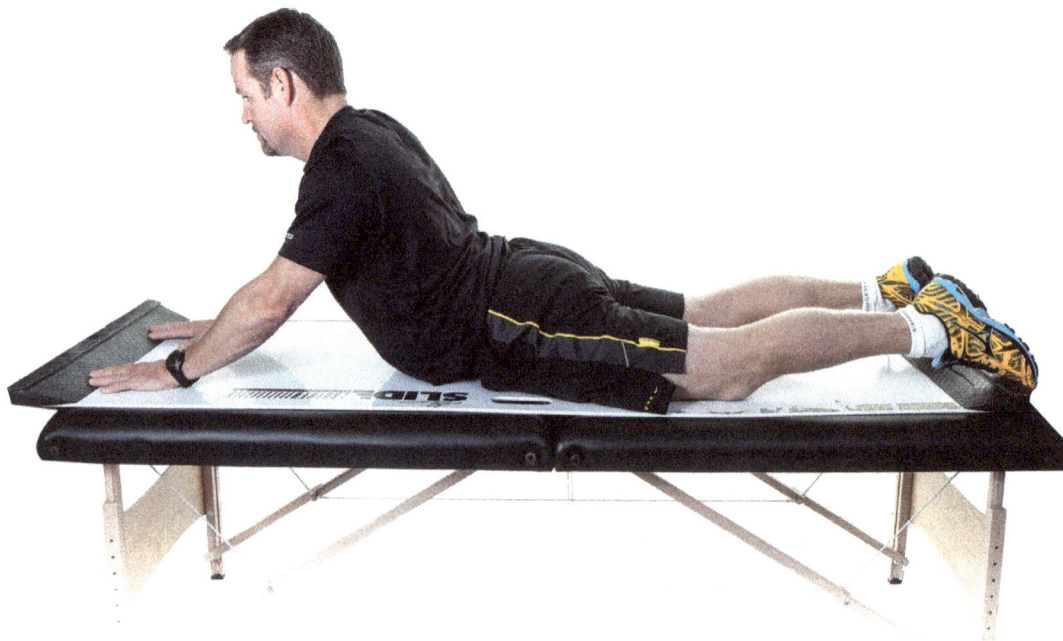

3. Inhale, then push from the hands and exhale as you straighten the arms to raise the chest off the floor. Do not go to pain and be alert to the possibility of pain in the back of the spine as the joints are being compressed. Keep pelvis and thighs in contact with the floor as you slowly extend the spine, moving from neck, to the upper, middle, and lower back. Try not to poke the chin forward and try to keep the shoulder blades back. You will feel this anywhere from the front of the neck through the front of the ribs and abdominals and down to the front of the pelvis.

Exhale as you press up. As you push to the top of movement, pull the chin straight back into the throat and pull the head back using the lower neck muscles. Look straight forward. Hold for 30-60 seconds while breathing normally, 2X as a static stretch, or repeat for 10 breathing cycles.

While the stretch is for the abdominals, it is very important to appreciate that you are also putting pressure on the facet joints in the back of the spine. This is very common and limiting in people with back issues. Due to the tendency for many of us to have tight hip flexors, lying on the stomach can cause the lower back to arch abruptly. Try to do the pelvic tilt and hold before attempting the press up. This takes the pressure off the lower vertebrae and distributes the pressure to the middle and upper lumber vertebrae and points higher up the spine. Keeping the shoulder blades back also helps to avoid the dreaded forward hump in the upper back.

# Middle and Upper back (Thoracic Spine)

## 13. Seated Torso Rotation

Equipment needed: Exercise mat, sturdy bench, or bed

The thoracic spine likes rotation, but is very dependent on rib movement. Cardiovascular exercise and deep breathing keep the ribs moving.

1. Sit with back in neutral position. Put one leg straight out and the other bent with the foot on the outside of the straight leg. Pull the top knee across the midline with the hand, which also stretches the outside of the hip in the buttock.

2. Maintaining good posture, slowly rotate toward side of the bent leg. Use the back of the opposite arm to push on the bent leg to enhance the stretch. Inhale and sit up a bit straighter, try to turn a bit more with each exhalation. Do not hold your breath. Hold the stretch for 30-60 seconds 2X, or repeat for 10 breath cycles.

Using the hand.

Using the elbow.

**CORE**

# 14. Extension Mobilizer

Equipment needed: Exercise ball, bench, or the arm of a couch

This mobilizer is designed to undo the hunched reality brought on by technology and our propensity to slouch while sitting or standing. This mobilizer opens the ribs to allow better thoracic extension and improved respiration. One cannot happen without the other.

1. Start by sitting upright on the floor in front of a bench turned sideways, with back against the edge of the bench. Clasp the fingers together behind the head. Inhale and hold the breath.

2. Holding your breath, attempt to bend back over the edge of the bench. You may feel a stretch in the front along the ribs. Hold for 5-10 seconds, then exhale and continue bending back further if possible. This will result in compression in the facet joints in the spinal column. Stay in this new position, and again inhale and give it a bit of release, hold again and go back further to stretch the front, exhale to go into a new position. Repeat 3 times.

3. Placing one hand on the lower part of the ribs will ensure that the required movement is not coming from the lower back. Try to not let the lower ribs flair.

Note: If you want to get to a particular point in the back, slide the buttocks forward or back to localize an area of your choice on the edge of the bench.

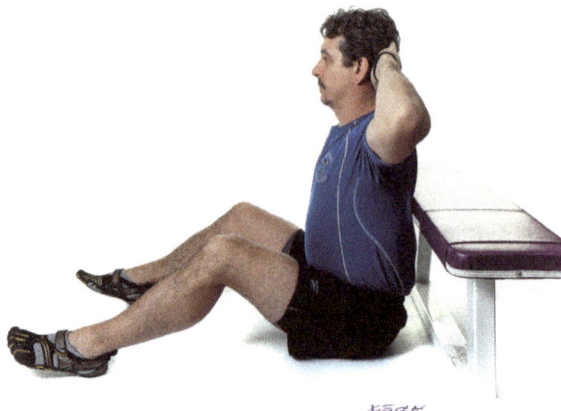

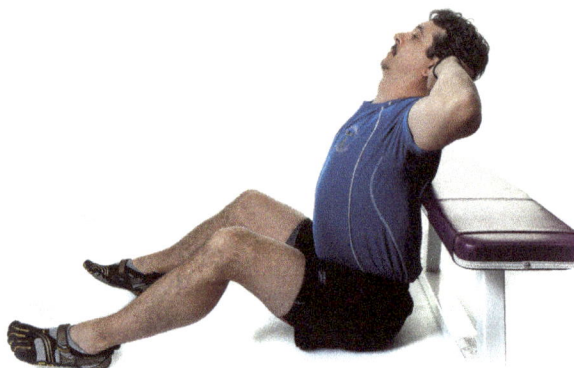

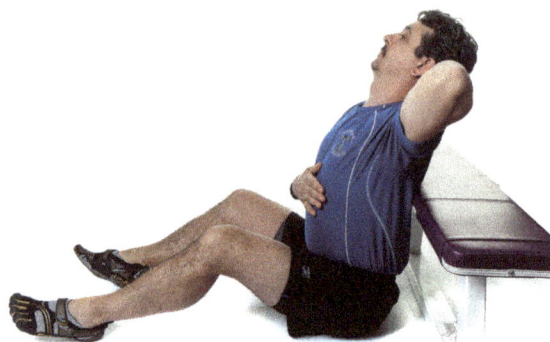

## 15. Cervical and Thoracic Spine Extension

Remember to take deep breaths throughout these dynamic mobilizers.

Equipment needed: Exercise ball

1. Sit on a ball then walk the feet out while laying back onto it, continue to roll out so that the back of the head can just touch the ball.

2. Now clasp fingers behind the head and let the elbows open toward the floor. This opens the upper chest and rib cage as well as mobilizes the upper thoracic spine with the help of gravity and deep breathing.

3. Hold for 30-60 seconds, or repeat 10X by raising and lowering the head. Inhale up and exhale down when using this stretch as a dynamic mobilizer. Roll toward the head to increase the range of motion.

**CORE**

## 16. Thoracic and Lumbar Spine Extension

This is a variation of the previous stretch, focusing on the lower back rather than the upper back and neck.

By walking the feet further out the shoulders will be brought further down the ball so we will get a deeper stretch lower down the spine. From this position, you are also able to bend the knees, and drop the hips which will effectively compress the lower thoracic and lumbar spine facet joints and stretch the abdominal muscles. Hold for 30-60 seconds as a static stretch, or repeat 10X as a dynamic mobilizer. Inhale up and exhale down.

## 17. Flexion Mobilizer

Equipment needed: Chair or bench

It is also important to maintain full flexion mobility of the entire spine. This stretch will keep the spinal extensor muscles both strong and supple to allow better posture. It is important to maintain flexion, but not quite as important as attaining and maintaining extension.

1. Begin in a seated position with the elbows on the knees and hands clasped. Keep the weight on the elbows throughout.

2. Start with a deep breath, and exhale as you bend head forward to look down toward the stomach while pushing the top of the pelvis and lower back toward the rear to round the entire spine. You will feel yourself rocking on the sitting bones as the top of the pelvis goes back.

3. From this position, take another deep breath, and on the exhale, attempt to flex further. Hold for 30-60 seconds, 2X, or repeat 10X as a dynamic mobilizer.

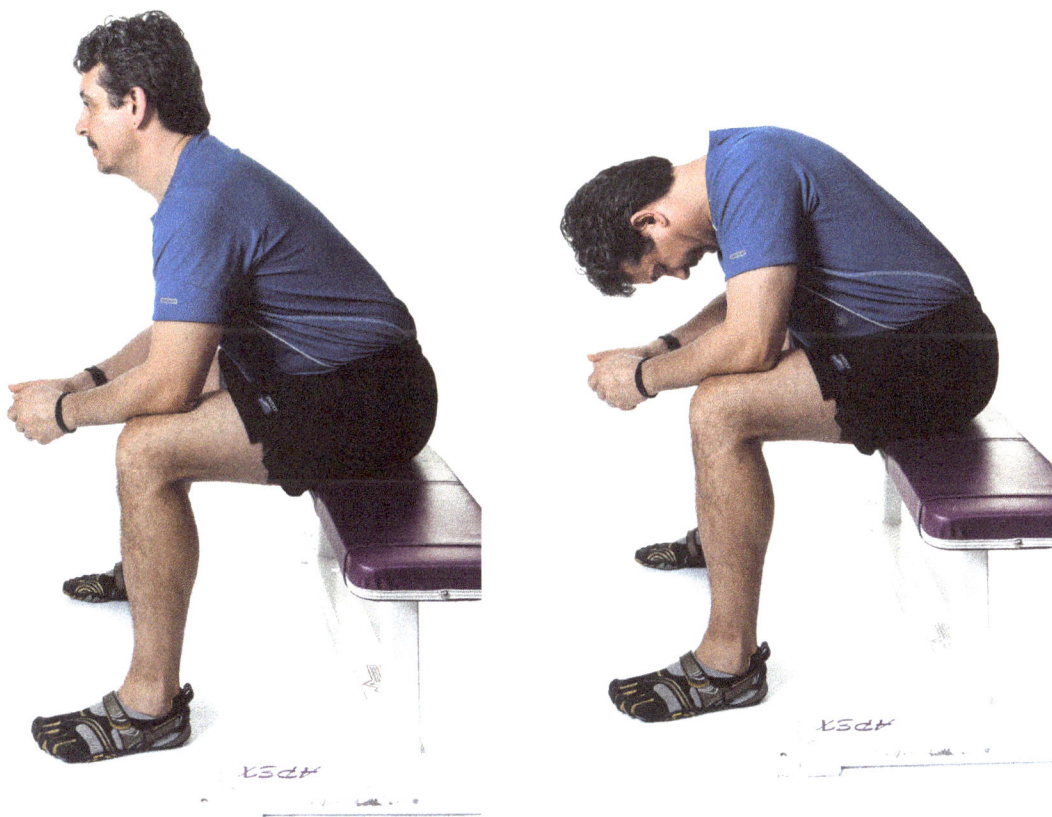

CORE

# Cervical Spine (Neck)

## 18. Retractions (Base of Skull)

As you sit in front of a computer and poke the chin forward, the muscles at the base of the skull can become very tight and this tightness can progress to causing or contributing to headaches.

Equipment needed: Chair or bench

### *Beginner*

1. Sit erect and place fingers on the chin with the heel of the hand on the chest. Brace the abdominals slightly to discourage arching the low back.

2. Pull the chin into the throat while guiding and applying a bit of pressure with your fingertips on the chin. Pull until you feel tension in the base of the skull or base of the neck. This tension can extend down into the middle back. This is a fantastic stretch to help alleviate tension headaches.

Hold for 30-60 seconds and take 4-6 deep breaths, or pull the chin in until you feel tension at the base of the skull, hold for 2 seconds, then relax to take the tension off, repeat 10 times. Do not push too hard or force the movement and try not to poke the chin forward on release. As you bring the chin back it may be difficult to swallow. Work toward being able to swallow with the head and neck in a retracted position.

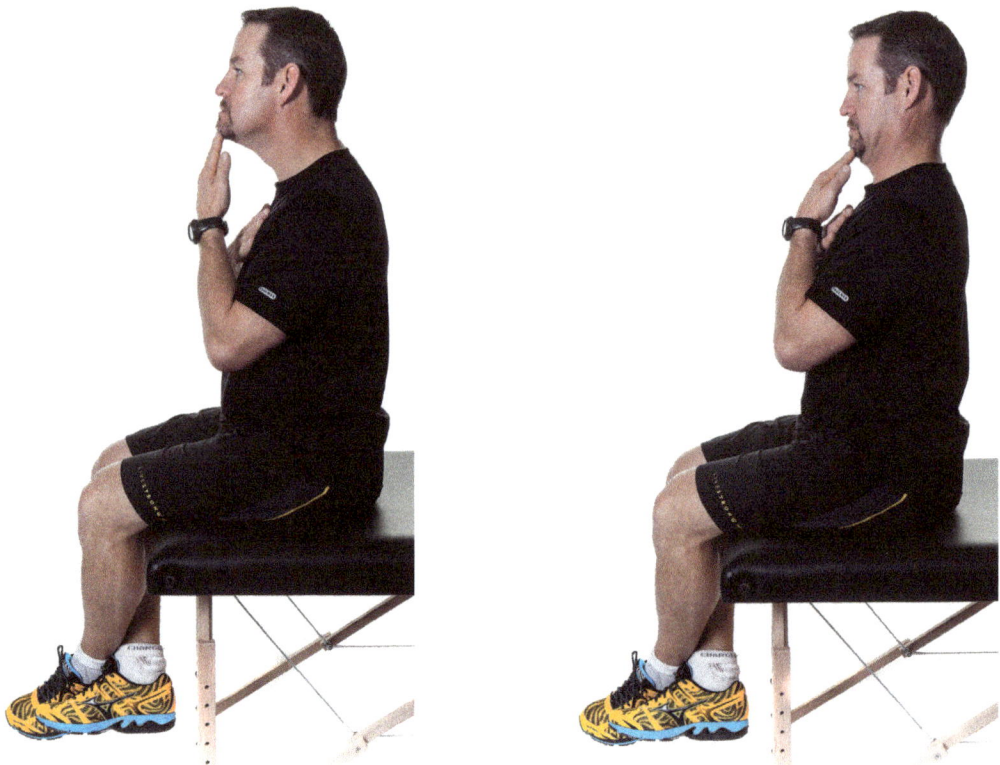

### Advanced

1. Start on your back with knees bent. Lift one knee then the other to 90 degrees.

2. Pull the knees into the chest. This will stop the lower back from arching and encourage more mobility through the middle back. Now pull the chin into the throat.

The stretch can be felt from the base of the skull down to the lower back, stretching the entire spinal musculature. Retractions will stretch the muscles at the base of the skull (suboccipital) and very similar to the set up for the lumbar flexion stretch (knees to chest), but more neck focused. Hold for 30-60 seconds, 2X, or repeat 10X with a 2 second hold as a dynamic mobilizer.

**CORE**

# 19. Neck Side Flexors/Upper Trapezius

Equipment needed: Chair or bench

Tip the head to one side and try to put the ear to the outside of the shoulder. The muscles on the opposite side of the neck are now working to stop the head from dropping to the side. Technically, this is not a stretch but rather a contraction of the upper trapezius muscle (the muscle that lifts your shoulder up to your ear) and the muscles on the side of the neck. Remember to do a static stretch—the muscles need to be relaxed.

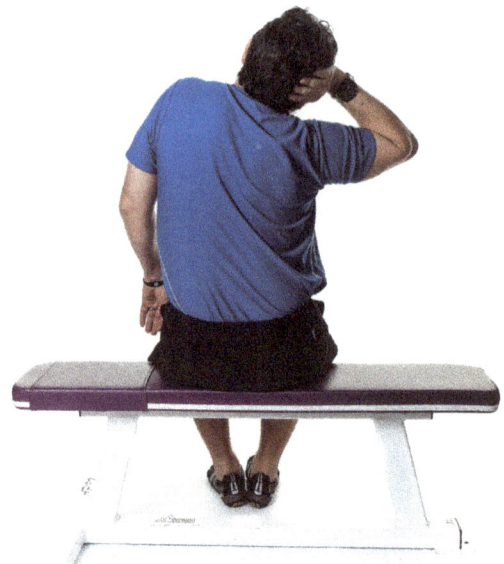

1. Place your hand high up on the same side of your head that you're tipping toward, and press the head into the hand which will enable the muscles on the opposite side to relax and able to be stretched. Let the neck and head go to the side as far as is comfortable and maintain support with the hand. This stretch includes the upper trapezius muscle and the deep muscles of the neck called the scalene muscles and the superficial muscle easily seen on the front of the neck called the sternocleidomastoid muscle.

2. By lifting the opposite shoulder in the direction you are moving, the trapezius muscle is now not influencing the neck muscles and the neck muscles are now able to be stretched further more specifically.

3. Control of the stretch is provided by the supporting hand on the head.

4. Once you are able to go as far as the neck will allow, push the shoulder on the opposite side down to stretch the upper trapezius muscle. *

*Placing the hand on a table or your hip to support the weight of the arm will allow the upper trapezius muscle that you are stretching to relax further. Lift this shoulder and try to move the neck to the other side again. At the end of the neck stretch, push the shoulder down again to further stretch the trapezius muscle. Hold each of the neck and trapezius muscles components separately for 30-60 seconds, and repeat 2 times on each side.

**CORE**

# 20. Neck Rotation/Side Flexion (Rotexion)

Equipment needed: Chair or bench

We have to turn our heads to look over our shoulder, to do a lane change, or look for something behind us. When throwing something or hitting something with a bat or club, we need to rotate the body below the head as we focus on what we are going to hit while keeping the head still. This requires the head and neck to tip to the side and rotate toward the back. This stretch is basically looking over the shoulder to look at the butt. If this hurts, our tendency is to move the head forward to stretch it out. This may feel better but hasn't fixed the problem of rotating and extending the neck, and may lead to the facet joints of the neck and upper back becoming weak and unable to take pressure. Putting pressure on the neck joints may feel uncomfortable, but with assistance to reduce this pressure, you can avoid pain and return to normal function.

*Beginner*

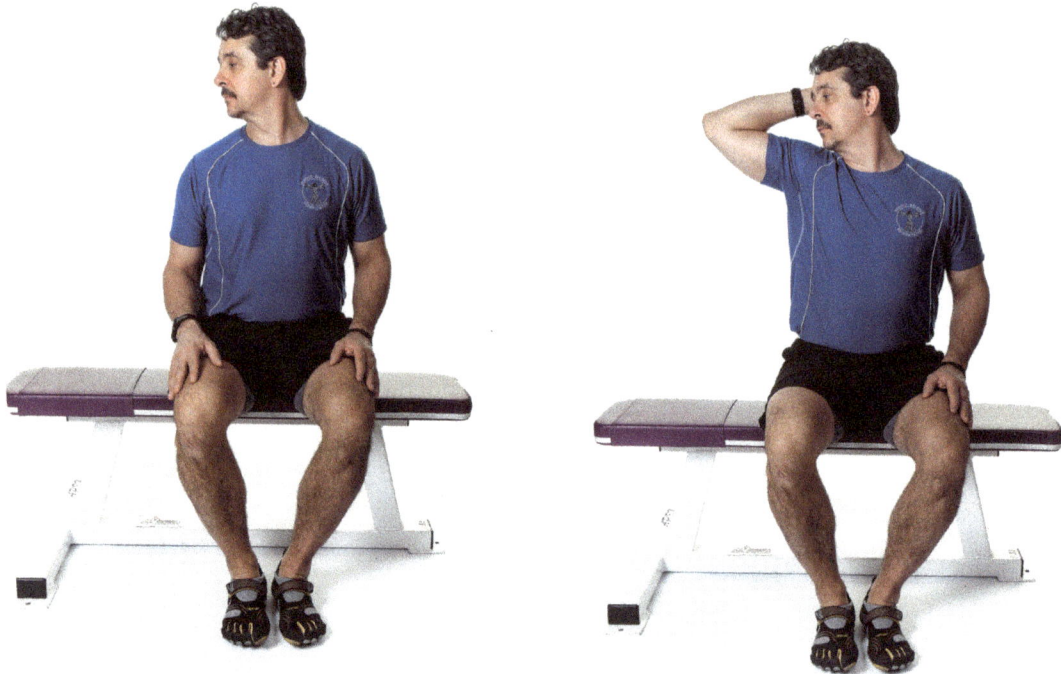

1. Turn the head to one side as far as possible.

2. Place the hand high on your head with your thumb just in front of the ear to form a window with the upper and lower arm to look through.

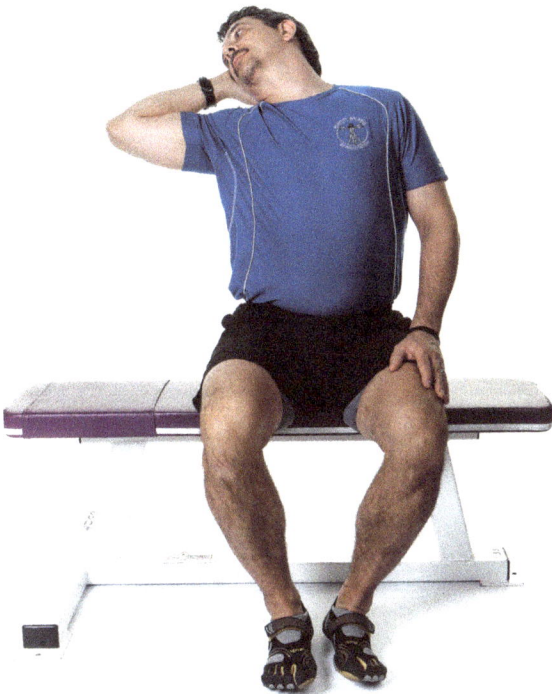

3. Press the head into the hand to support its weight as you allow the head to go back so that the ear is trying to be placed over and toward the back of the shoulder. You are compressing the neck joints on the same side you're going toward, as well as stretching the sternocleidomastoid and scalene muscles on the front/opposite side of the neck. Provide enough resistance on the support hand to prevent any pain in the back of the neck and direct the eyes toward the buttocks. Putting the middle and lower back against a high-back chair will help to avoid a lower back arch, and isolate movement to the neck.

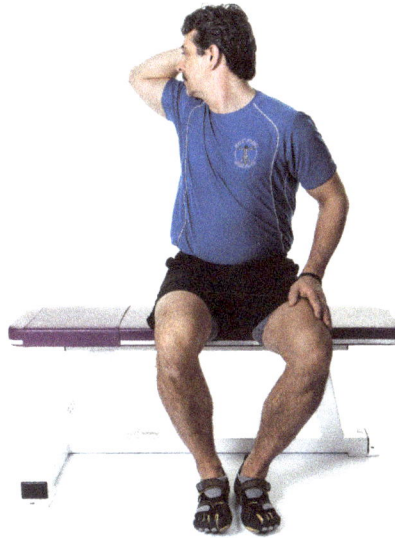

4. Bring the head back up with the hand but keep the head in the rotated position. Turn the head further and repeat the "turn and tip" 3 times to each side. Full rotation will eventually have your chin on your shoulder.

5. Placing a hand on lower ribs will help detect arching of the lower back. Try to keep the chin down and look toward your buttocks.

**CORE**

Rotexion is a stretch that will prevent the dreaded neck hump, and may even help to stop and potentially reverse this process. Instead of going forward to "stretch it out," go against the effects of gravity and go toward the back.

### Advanced

With practice, you will be able to turn and tip your head more easily and you will need less assistance with the hand. Rotate the head, then side flex the neck to look over the shoulder without assistance while placing a hand on the lower ribs to provide awareness of arching the lower back.

### Standing Advanced (progress check)

While standing, along with the standing extension stretch (page 11), reach back to touch the inside of the knee and look over the shoulder at the same time to involve the entire spine.

This would be a dynamic stretch, but can be used to help identify restrictions requiring more specific static stretching, and in which area; cervical, thoracic, or lumbar region.

# Lower Body

## Front of Hip (Hip Flexors)

### Psoas Major/Minor and Iliacus

The hip flexors connect the legs to the pelvis as well as directly to the spine. As the name indicates, they act to either lift the thigh toward the trunk or the trunk toward the thigh. The hip flexors also play a crucial role in controlling the angle of the pelvis relative to the lower spine and stretching these muscles as well as strengthening the lower abdominal muscles can help to eliminate a sway back.

Anyone who spends long periods of time sitting (most North Americans), will experience a shortening of these muscles. This is readily seen as we walk like old people after getting up from prolonged sitting such as a long drive. Mastering this stretch will change how you walk and run, how you carry yourself, and the hip flexor stretch is part of a comprehensive back care program to help eliminate back pain.

We feel strongly that lengthening these muscles is one of the most important things you can do for your spine, pelvis and legs.

*Test Yourself*
Lie flat on your back with legs straight on the floor, toes pointing at the ceiling. Notice any tension in the lower back? Now bend your knees, and if bending your knees decreases low back tension you have tight hip flexors and you need these stretches every day.

## 21. Standing Hip Flexor Stretch

Equipment needed: Door frame, outside corner of a wall, or the leg of a squat rack

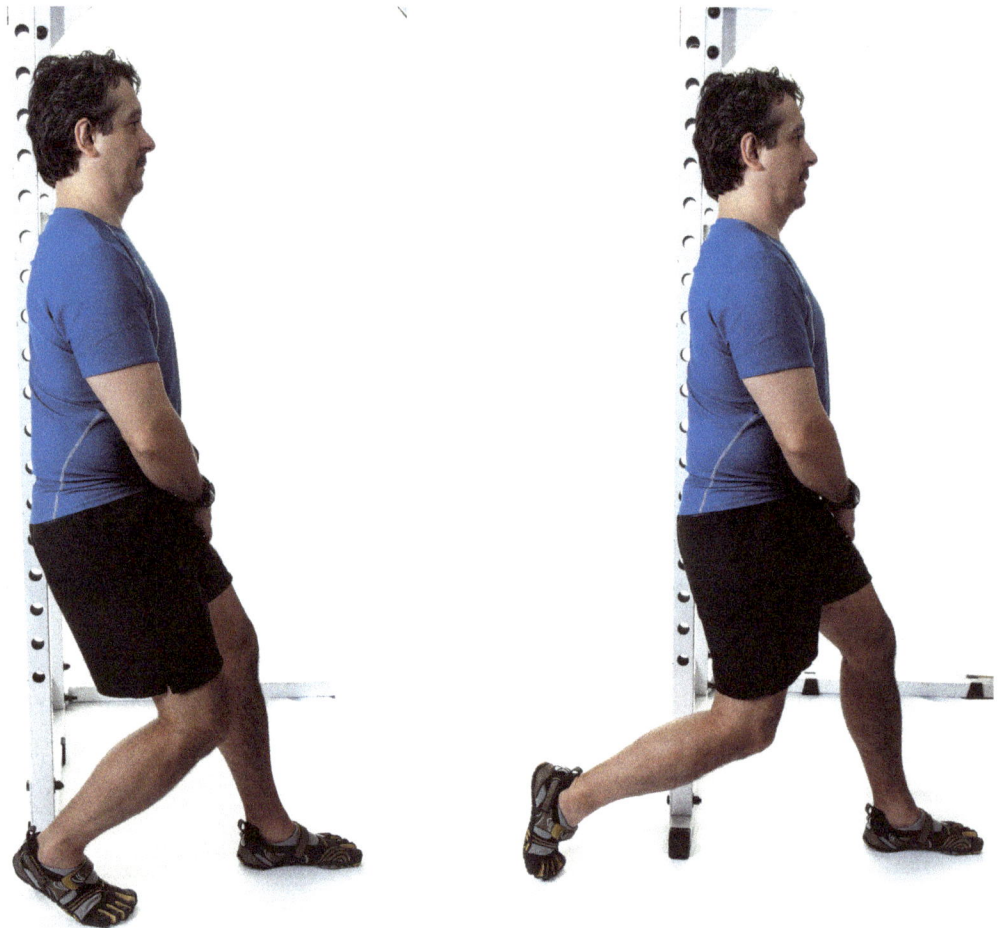

1. Stand with the middle of the back flat against a wall near the edge of an outfacing corner or the leg of a squat rack so that you are able to put one leg behind you and the other leg to the front. Standing with the middle of the back against one side of a door frame and the other foot straight ahead on the other side of the door frame is an ideal place to do this stretch. The hands can also be used to push on the door frame in front of you to help keep the back flat.

2. Assume a staggered stance position, placing the lower leg and foot being stretched to the rear so the knee is directly below the hip and the rear foot is straight up on bent toes. The rear foot and thigh are both straight up and down. The front leg is slightly bent and supporting your weight. Be sure not to bend the front knee too much.

3. Engage the abdominals and tighten the buttocks to flatten the lower back against the wall in a pelvic tilt. Only use enough pressure on the back leg and foot to control the leg while you attempt to straighten the leg at the knee and hip until a comfortable stretch is felt in front of the targeted hip. Maintain contact of the lower back with the wall. Try coughing to engage the core or place your fingers behind the lower back, palm side on the flat surface, and press your back into the fingers. If the back arches, you've lost the pelvic tilt and the hip flexors will remain shortened rather than lengthening. You are not trying to put the heel to the ground but rather are pushing the knee back in an arc of motion while pivoting on the toes. If you are able to contact the heel you will need to re-establish the start position of the rear foot by placing the toes back a bit further.

**LOWER BODY**

If it is difficult to keep the back flat, please refer to core stretches and practice the laying (p 15) and standing pelvic tilt (p 16) until you can take that same movement and apply it to this standing hip flexor stretch. Without it, you will miss lengthening this chronically short and tight muscle group. The standing hip flexor stretch could be considered the gold standard of a static hip flexor stretch. Hold for 30-60 seconds, 2X as a static stretch, or repeat 10X as a dynamic mobilizer.

## 22. Kneeling Hip Flexor Stretch

Equipment needed: Exercise mat, floor, and an exercise ball or chair

From staggered half-kneeling position, keep the front knee positioned behind the toes. Establish and maintain a posterior pelvic tilt, then slide your rear knee back until comfortable tension is felt in front of the hip. Control is maintained by putting pressure on the front leg and by placing the hands on a ball or chair and easing forward to increase tension and backward to release tension. Hold for 30-60 seconds 2X as a static stretch, or repeat 10X as a dynamic mobilizer.

**LOWER BODY**

## 23. Standing Dynamic Hip Flexor Mobilizer

For those who have mastered the standing static stretch against a flat surface, establish a pelvic tilt in the staggered stance and then lengthen through the front of the hip by pressing the heel away. To activate the hip flexor muscles even more, incorporate a bounce into this dynamic stretch. Standing up at the desk after long periods, on the first tee, standing while fueling up during long drives. This is a dynamic stretch that will keep the range of motion you have and loosen you up a bit. The preceding static stretches are the best way to actually increase the available range of motion to function at your best.

# FABERS (Flexion, Abduction, External Rotation) ("Figure 4 stretch")

### 24. Front of the Hip Joint (Anterior Hip Joint Capsule)

Equipment needed: Exercise mat or carpeted floor

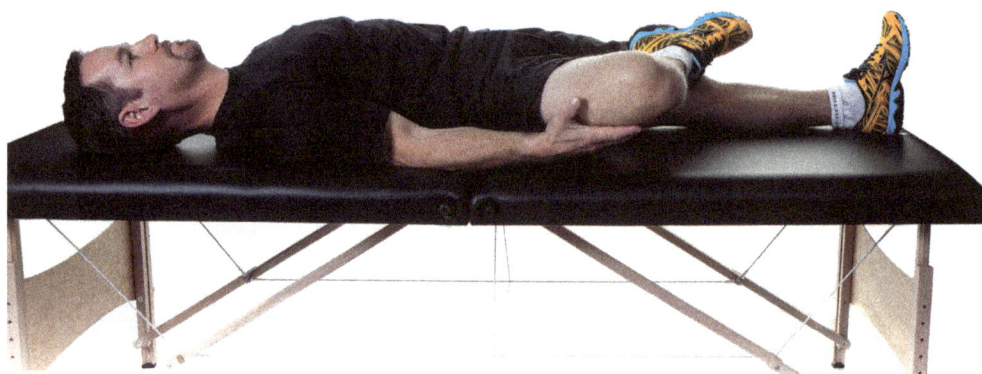

1. Lie flat with the ankle on top of the other leg, either above or below the knee cap.
2. Let the knee down to the side while placing a hand under the knee to assist and stop the muscles in the front of the hip from holding.
3. Activate the buttock muscle on the same side to help pull the knee down to stretch the front of the hip joint capsule and the short adductors (muscles of the inner thigh).

Hold for 30-60 seconds 2X as a static stretch, or repeat 10X as a dynamic mobilizer.

## Buttocks (Gluteal Muscles)

### Gluteus Minimus/Medius and Maximus

Equipment needed: Bench, bed, or sturdy shorter table/coffee table, or an exercise mat to do the lying stretch

The gluteal muscles (glutes) are referred to as the "wheelhouse". They are the largest muscle group on the human body—they are meant to be the power centre. In general, North Americans are modest to a fault and don't talk about, use, or move the gluteal muscles as they are supposed to be used. They function to extend (toward the back), abduct (to the outside) and rotate the thigh to the outside as well as to control the pelvis by pulling the back of the pelvis down toward the heels.

## Seated Gluteal Stretch

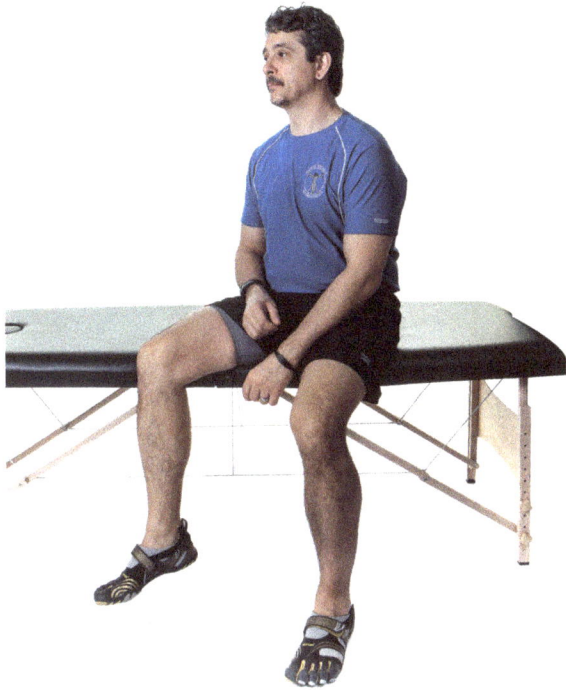

1. Sit on a bench with the hip bent to 90 degrees.

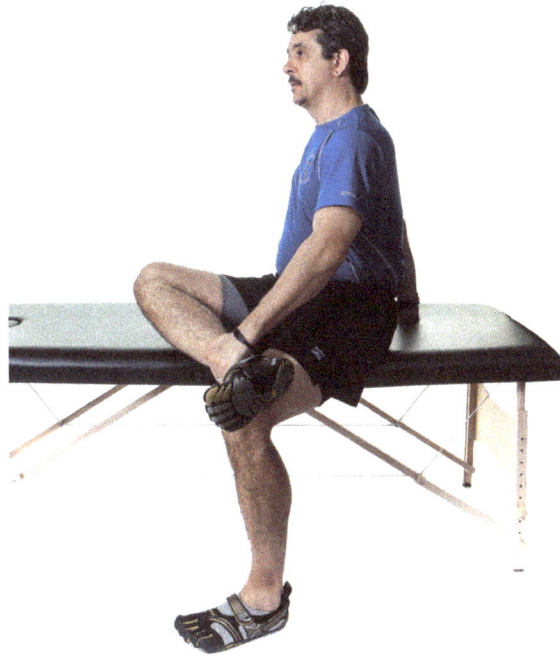

2. Lift the knee on the bench.

3. Turn the knee and body toward the front as much and as far as is comfortable and put the foot on the edge of the bench if possible. Ideally you want to have your thigh straight out in front of you (perpendicular to the bench). Do not force this stretch, back off if you have pain and allow the foot to be off the edge of the bench initially. Ensure the hips are in line with the bench as much as possible (parallel) and the back leg is only used to maintain balance. Do not push the back leg straight as this would open the hips and you will miss the gluteus minimus and medius portion of the muscle. This is the reason that doing this on the floor is less effective.

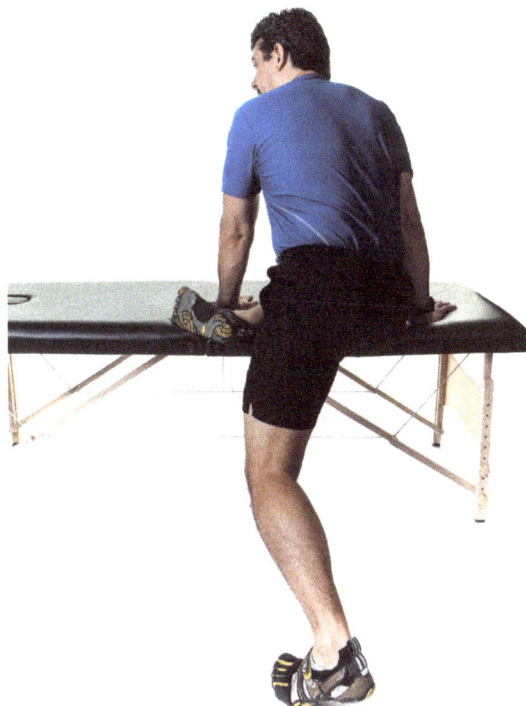

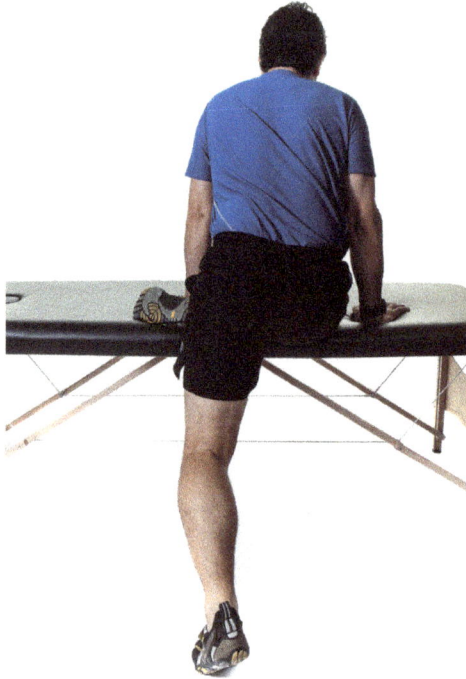

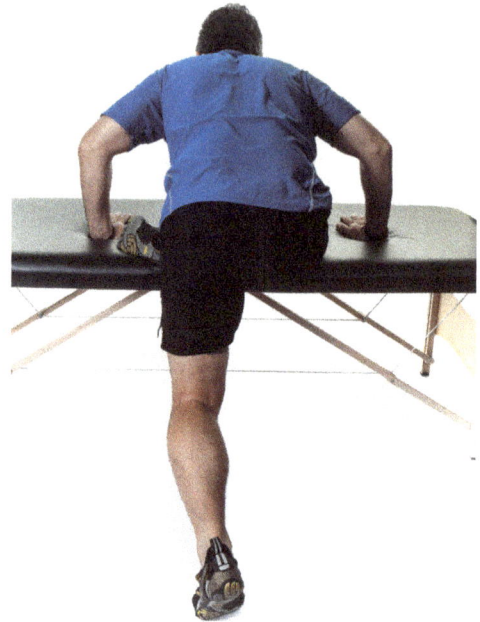

4. Straighten the back and place the hands onto bench on either side of the knee. The upper 1/3 of the thigh on the leg being stretched should be back and off the edge of the bench to enable turning the opposite hip forward.

5. Turn the opposite hip forward so the hips are parallel and the thigh is perpendicular to the bench. You may feel tension at the top of the hip on the outside in the smaller gluteal muscles (gluteus minimus and medius).

6. Keep the back straight and lean forward on to the hands with them positioned on either side of the knee. Keep the centre of chest over the knee. Use the arms to control your descent and a stretch will be felt further down the back outside of the thigh. If you are unable to bring the knee and thigh all the way up on the bench to a perpendicular position, only go as far as comfortable and do what you can.

Hold for 30-60 seconds 2X as a static stretch, or repeat 10X as a dynamic mobilizer.

Lifting the hands would cause the glutes to contract, and this would no longer be a static stretch.

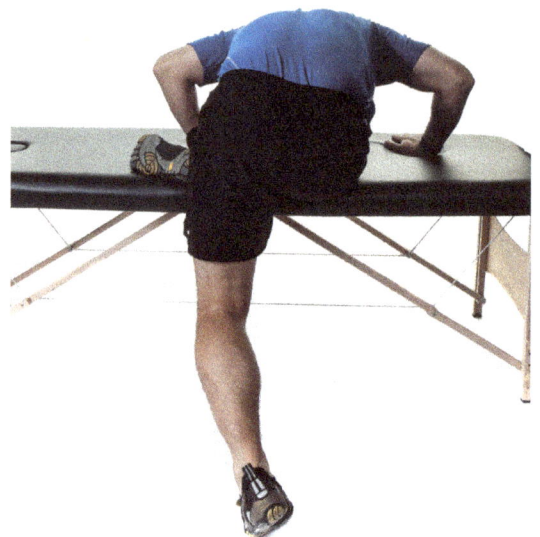

## 25.  Laying Gluteal/Piriformis Stretch

1. Lay flat on your back with both knees bent.

2. Lift one leg and place the ankle across the top of the other thigh near the upper knee.

3. Lift the foot of this leg off the floor and reach both hands around the thigh and clasp your fingers together behind the lower thigh. If you are unable to reach the leg comfortably, leave the hands out of the equation.

4. Slowly draw both knees toward the chest. Use a bench or chair to support the non-stretched leg once you are in position.

5. Gently pull the legs toward you until you feel the stretch deep in the buttock (this is where the piriformis muscle resides).

Hold for 30-60 seconds 2X as a static stretch, or repeat 10X as a dynamic mobilizer.

This stretch is often prescribed to sciatica sufferers. The piriformis and sciatic nerve are intimately related at the back of the hip and when the piriformis is bound up, tight and short, gentle lengthening can often provide sciatic relief.

(For more info on Sciatica and a great long- term fix, see Sciatic Nerve Tensioner Stretch on page 51.)

**LOWER BODY**

# Front of The Thigh (Quadriceps)

Equipment needed: Anything you can use to maintain balance while standing or an exercise mat when lying down

The quadriceps muscle (quad) extends the lower leg, as in when you are kicking a ball or slowing down or stopping your forward motion when walking or running. One part of this muscle group also assists in hip flexion or lifting the thigh to the chest. These are great stretches for the front of the thigh and provide relief for patellar (knee cap) pain or tightness.

## 26. Standing Quads Stretch

**LOWER BODY**

1. Bend the knee and pull the heel toward the buttocks by grasping the foot just below the ankle. Perform a pelvic tilt and you may already feel a tug in the front of the thigh.

2. Bend the supporting leg slightly at the knee. Maintain a pelvic tilt as you allow the foot of the leg being stretched to go back away from the buttocks extending from the hip, reaching the foot back as far as possible. Stop if you feel the lower back start to arch.

3. Now pull heel toward buttock as much as able until mild tension is felt in the front of the thigh. As you are also stretching the femoral nerve, the sensation of a sting may be noticed as well. Do not pull into pain. It is important to note that as you look down, the knee should be in line or better yet, behind the line of the hip.

Hold for 30-60 seconds 2X as a static stretch, or repeat 10X as a dynamic mobilizer.

## 27. Laying Quad Stretch

1. Lying on the stomach, grasp the foot and pull your heel toward same side buttock. To increase the stretch, pelvic tilt, lifting the top/front of the pelvis off the mat and tuck the buttocks down toward the heels. Try to flatten the lower back by contracting the gluteal muscles and abdominals together. Lying on the stomach helps to discourage the back from arching.

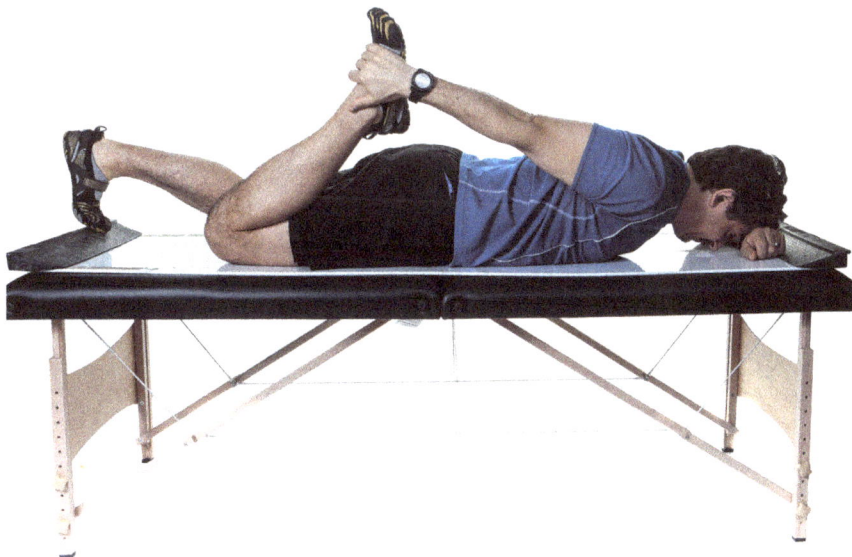

2. If you struggle to reach your ankle without sending the hamstring or lower back into spasm, use a skipping rope or towel to wrap around the ankle and then gently coax the ankle toward the buttocks.

Hold for 30-60 seconds 2X as a static stretch, or repeat 10X as a dynamic mobilizer.

# Back of The Thigh (Hamstrings)

Equipment needed: Bench, chair, or stairs, and an exercise mat

The hamstrings act to flex the knee, as when you kick yourself in the buttocks, and also to extend the hip when you stand from sitting or when you go down or come up from grabbing something off the floor. They also help control various rotations and counter-rotations at the hip, knee, and ankle while walking or running.

They tend to shorten when we spend most of our time sitting.

If you try to touch the toes, but can't quite get there and hold this position, is this a stretch? The answer is NO. The tension you feel is the muscle contracting to prevent you from falling over.

To stretch, we must support the trunk and upper body weight with the hands to allow the muscle to relax.

## 28. Standing Hamstrings (Advanced)

Equipment needed: Chair, bench, or couch

1. Hinge forward from the hips and place your hands on a bench, chair, or stair to control how far you are able to bend. Maintain a straight back and bend the knees slightly while holding the stretch. Increase the stretch by slowly lowering the upper body from the hips while controlling

the descent and taking the load off the hamstrings using the arms. The legs remain in position. Imagine lifting the tailbone up to the ceiling without straightening the legs. You should feel a stretch in the upper portion of the hamstrings. If you slowly straighten the knee slightly, you will now stretch the lower portion of the hamstrings. If you feel a sting in the back of the knee, bend the knee until the sting is gone.

Hold for 30-60 seconds 2X as a static stretch, or repeat 10X as a dynamic mobilizer.

2. Sciatic nerve stretch: Performing this stretch with a straight knee will result in a stretch of the sciatic nerve and a sting behind the knee. The nerve is highly sensitive and you can lengthen the nerve, but be cautious.

   This is only a beginner stretch to help relieve some sciatic nerve issues. Try to lift the front of one foot and the toes off the floor when in the stretch position and you can increase the tension of this nerve. This will help you to appreciate the difference between stretching a muscle and stretching a nerve.

Hold for 30-60 seconds 2X as a static stretch, or repeat 10X as a dynamic mobilizer.

## 29. Laying Hamstrings (Beginner)

The application for this position is for the novice stretcher, to teach neutral spine and ensure that the movement comes from the leg and not the spine.

1. Lay flat on the back in perfect postural alignment, with the neck and lower back as flat as possible. Raise the upper leg so the knee points at the ceiling and interlock the fingers behind the lower thigh, then slowly extend the lower leg to the first sign of tension. Point the toe to the ceiling to minimize sciatic nerve tension.

2. Pull the thigh tighter to the chest while keeping the same bend in the knee to stretch the upper portion of the hamstring further. Keep pulling the thigh and hold tight then slowly straighten the knee to stretch the lower portion of the hamstring.

Hold for 30-60 seconds 2X as a static stretch, or repeat 10X as a dynamic mobilizer.

# 30. Sciatic Nerve Tensioner/Stretch

Equipment needed: Stairs, leg of chair or table

The sciatic nerve tensioner is used for sciatic pain or numbness, as well as relieving the "nerve feeling" into the legs or feet.

1. Place the foot on the leg of a bench or chair. Slowly lift the toes up while keeping the heel down to get a light stretch in the calf. This can also be done by placing the ball of the foot on the edge of the bottom stair to get the initial tension in the calf. Slide the foot down to contact the heel to the ground when using the stair, if possible.

2. Put your hands on the bench, chair seat or stairs, and lean forward at the hips and allow the spine to curve. Keep the weight on the hands as you bend the elbows to control the descent while keeping the head up. Tension may be felt in the back of the calf, knee or even up to the thigh and buttocks. This is the sciatic nerve and is easily felt between the hamstring tendons behind the knee. It will feel like a cord that is extremely sensitive. Don't push to the point of pain.

3. Reach the head down to touch the chin to the chest, but go slowly. You may feel tension increasing in the back of the thigh, buttock, or up into the back. Don't push to the point of pain.

4. Lift the head and try to bend at the hips by bending the elbows more to increase the stretch. Then touch the chin to the chest again. Moving the head up and down will control the intensity of the stretch. Bending and straightening the knee or moving the heel up and down will also affect the amount of stretch tension. Do only one area at a time, and try to determine which area needs work. Always maintain control.

Hold for 30-60 seconds 2X as a static stretch, or repeat 10X as a dynamic mobilizer.

LOWER BODY

## 31.  Outside of Hip (Tensor Fascia Latae (TFL)/ Gluteus Minimus and Medius)

Equipment needed: Bench or cupboard

These muscles can get tight from walking with duck foot position (toes pointing outward) due to a previous injury to the ankle. You can get symptoms in the knee or down the outside of the leg or hip. When tight, these muscles can make a cross-over to the affected side difficult. These muscles can get injured when slowing down and cutting away from the planted leg. This stretch gets to the possible root of the problem.

*Beginner*

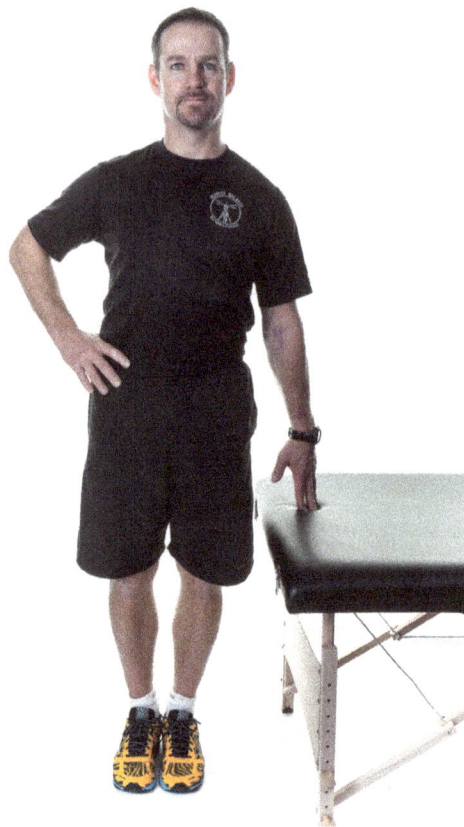

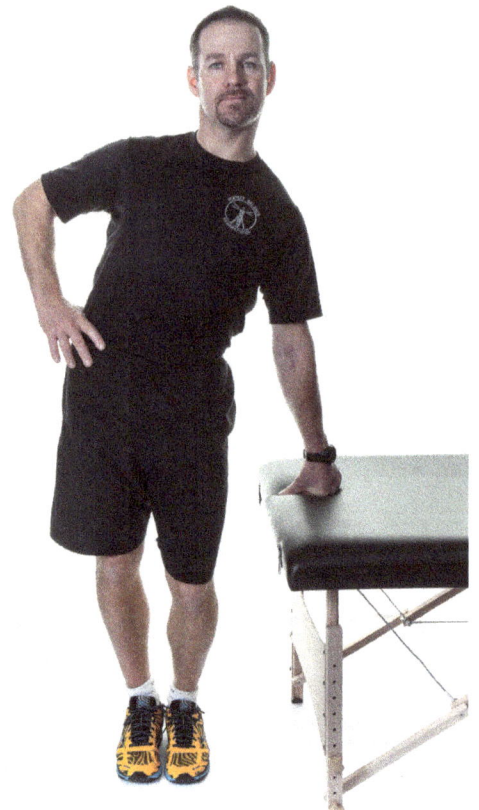

1. Stand against a high bench or cupboard. Put a hand on the bench to support the weight of the body. A cupboard is preferred as it is more stable than a bench and won't slip away.

2. Bend the knee nearest the bench while keeping the outer leg straight. Support your weight with the hand and the bent leg on the same side to control the stretch of the upper hip on the opposite side. You should feel tension in this area including the front part of the gluteal muscles. DO NOT let the lower back arch—keep the spine in neutral position. Hold for 30-60 seconds, 2X each side, or repeat 10X as a dynamic mobilizer.

LOWER BODY

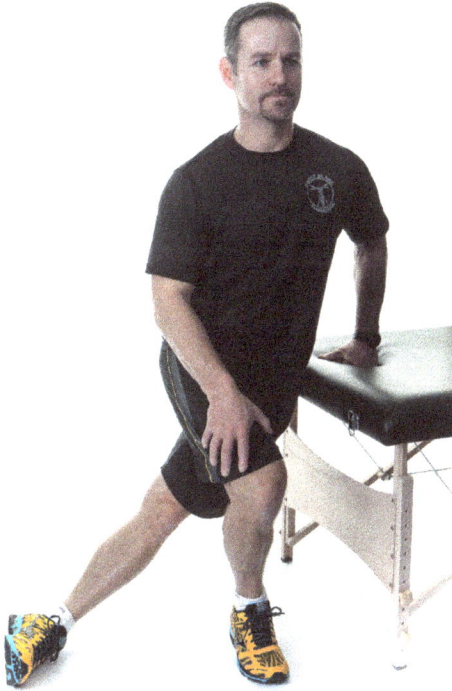

1. Place a hand on a high bench and press the outside of that hip against the bench. A cupboard is preferred as it is more stable and you can put your elbow on the cupboard for better control. Reach the inside leg behind the bent supporting leg and place your same-side hand on the thigh. When the hip and ribs are against the cupboard the movement will come from the hip and not the spine. The outside of the hip on the back leg is being stretched, so try to minimize the pressure on it. Make sure to keep the spine straight.

2. Place very little weight on the back leg. Allow the foot of the back leg to roll inward at the ankle and try to lift the outside of this foot off the floor to feel what no pressure feels like, then bend the supporting leg and apply just enough pressure to affect a stretch. Have the back leg straight out to the side. Maintain a pelvic tilt.

Placing the rear leg further backward will have more of an effect on the TFL. Placing the rear leg further forward will hit the glute medius and minimus more effectively. Hold for 30-60 seconds, 2X each side, or repeat 10X as a dynamic mobilizer.

# Inner Thigh (Adductors)

Equipment needed: Exercise mat, bench, or cupboard

The adductors are the muscles that bring the legs toward the middle of the body. As infants, our legs are wide apart and turned outward and the feet are flat like little pancakes. As we develop balance and strength in the adductors, the legs start to get closer and our arches start to develop. Our whole life is spent trying to keep our legs relatively close and gravity spends its time trying to bring them right back to the outside. You'll notice that most seniors have a wider stance and sort of wobble as they widen the base of support when walking. The adductors also function to help us change direction and pull our bodies to the outside as we bring one leg across in front or behind the other leg.

## 32. Standing Groin Stretch (Long Adductors)

The adductor muscles that cross both the hip and knee joint require a straight knee to stretch the muscle properly.

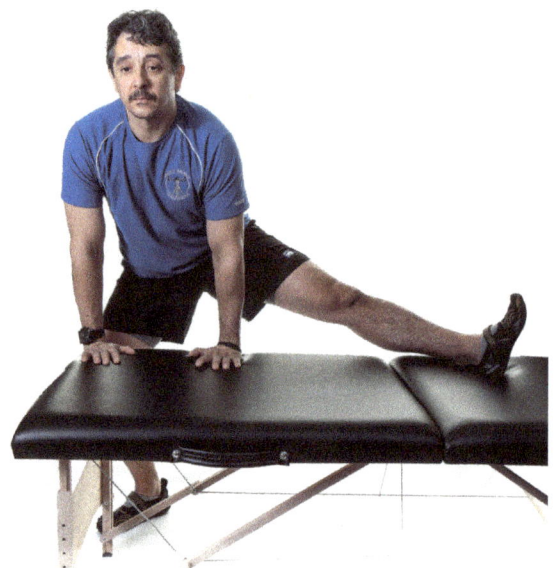

1. Place the inside of your foot out to the side on a bench and put your hands on the bench so you feel balanced. Bend the knee of the support leg to control the intensity of the stretch and lower slowly until you feel a comfortable stretch on inside of the thigh. Keep weight on the supporting leg and hands to help keep the muscles relaxed and not engaged. With the inside of the foot flat on the bench this stretch is adductor-specific.

2. Stand tall and back out of the above stretch, and then turn the toe toward the ceiling and slowly ease back down into the stretch. This moves the line of pull down into the inside of the back part of the thigh or the inner hamstring muscle. These will usually feel tight sooner than the pure adductors. Hold for 30-60 seconds, 2X each side, or bounce a bit as a dynamic stretch and repeat 5-10 times.

## 33.  Seated Groin or Butterfly Stretch (Short Adductors)

Sit straight or lean against a wall with the knees bent, bottoms of the feet placed together and knees wide apart.

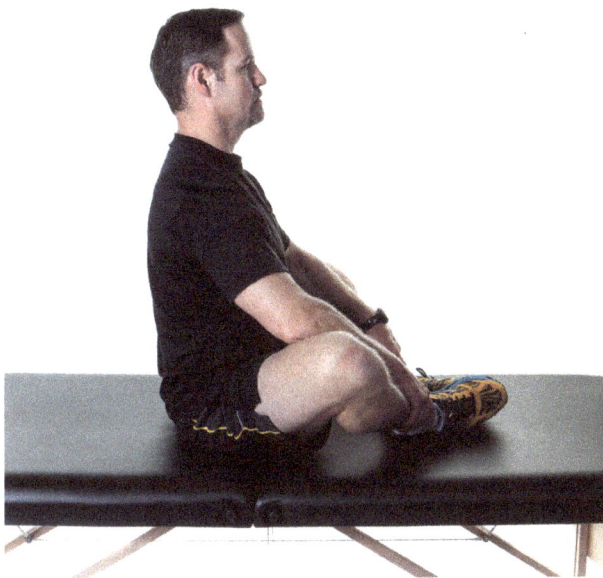

Increase the stretch by leaning forward with a straight back and opening the thighs with your elbows or pull the heels closer to groin. Hold onto the ankles without pulling up on toes and press elbows down on thighs to add to the stretch if the knees can get almost to the floor.

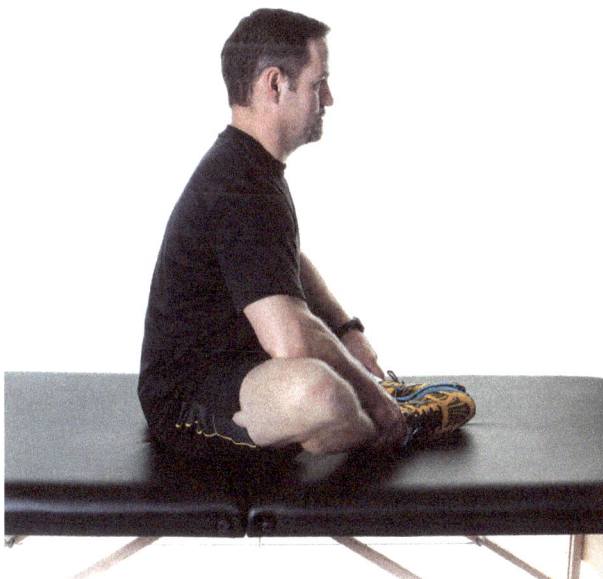

*Beginner*
Try to pull the knees down by engaging the gluteal muscles while placing the hands UNDER the knees to control the descent of the legs. The weight of the legs keeps the muscles engaged and pushing the knees down into the hands will allow the "short" adductors to relax.

# Lower Leg (Calves, Gastrocnemius and Soleus)

Equipment needed: Wall or legs of a squat rack

Do you recall the secret to effective static stretching? If you stand with the toes on a stair and drop the heels, is this a static stretch? No, this would be considered an isometric contraction. You are feeling the muscle working, so it is not relaxed and stretching.

## 34. Upper Portion, Back of Lower Leg (Gastrocnemius (Gastrocs))

The gastrocs begin above the knee then join the soleus to become the Achilles tendon at the heel. These muscles propel us forward in running. Many people refer to this stretch as the "push the wall down stretch."

Important Note: Don't push with the front of the foot down into the floor as the muscle we are trying to lengthen will then be contracting. As with all static stretches, we must relax the muscle for it to lengthen.

Place the hands on a wall with a staggered stance targeting the back leg. Lock the rear knee and turn THE LEG outward without rotating the pelvis or foot. (You should feel the inside arch of the foot lift as the ankle turns outward).

Three rules to apply; keep knee locked, foot stays pointed straight, and heel remains on the floor.

Slowly lean forward until you feel a very slight stretch in the calf of the rear lower leg. Ensure that the centre of the knee is aligned with the centre of the foot. Take the weight of the body on the front foot and the hands. Apply only enough pressure in the back leg to elicit a stretch. Keep the core straight but not rigid. Hold for 30-60 seconds 2X, or repeat 10X as a dynamic mobilizer.

## 35. Outside of Lower Leg (Peroneals (Longus and Brevis))

These muscles run down the outside of the lower leg and flex to propel us forward and sideways toward the inside of the foot, (cutting laterally on a basketball court). The peroneals can be injured by an inversion sprain where the foot rolls inward spraining the ligaments on the outside of the ankle at the same time as injuring the peroneal muscles.

Start in normal calf stretch position. Turn the knee and ankle outward. Rotate the leg, keeping the foot straight and the knee locked. Step forward and to the opposite side with the other foot. Lean forward to where you can see the outside of the hip lined up over the big toe as you look down under your arm past the outside of the hip on the leg being stretched.

You should feel the stretch on the outside of the lower leg, the back of the ankle, into the outside lower part of the ankle or the outside of the foot.

Hold for 30-60 seconds, 2X each side, or repeat 10X as a dynamic mobilizer.

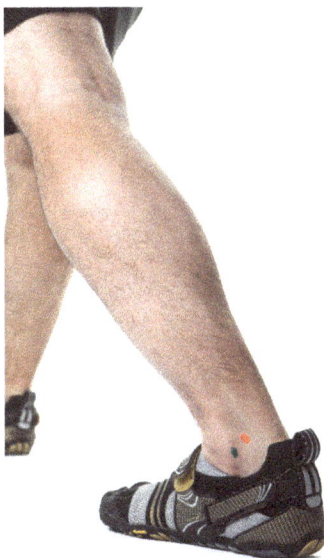

Leg rotated inward, arch falls
**Blue dot:** Ankle Bone

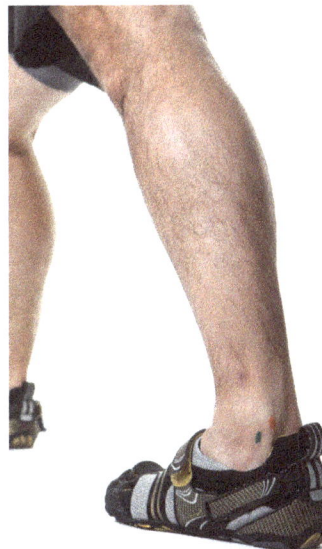

Leg rotated outward, arch lifts
**Red dot:** Peroneal Tendons

## 36. Inside of Lower Leg (Tibialis Posterior)

This muscle runs up the inside of the lower leg and works to propel the leg powerfully inward as we cut outside, and it helps control our gait. This is a great stretch for posterior shin splints, that annoying area on the inside, back of the shin.

Start in the normal calf stretch position. Lock the rear knee and cross the other leg in front so that you can see the outside of the opposite hip on the outside of the foot toward the little toe. The knee of the stretched leg remains straight.

You should feel a pull on the inside of the lower leg or into the inside back of the ankle and down to the inside of the foot.

Support your weight on the hands and front leg, putting as little weight as possible on back leg while maintaining the straight position of the foot. Hold for 30-60 seconds, 2X each side, or repeat 10X as a dynamic mobilizer.

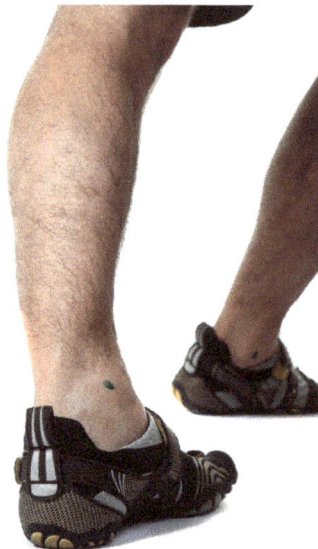

Leg rotated inward, arch falls
**Blue Dot:** Ankle bone

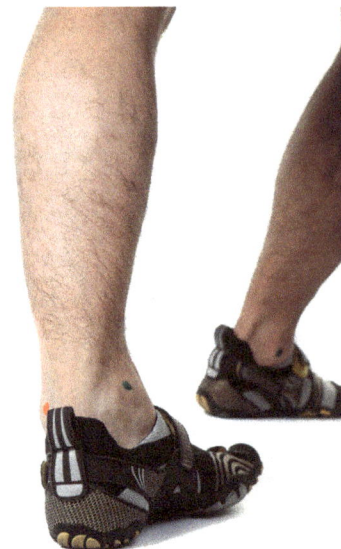

Leg rotated outward, arch lifts
**Red Dot:** Outside Ankle bone

## 37. Lower Portion, Back of Lower Leg (Soleus)

This muscle does not cross the knee joint so we need to bend the knee to stretch soleus.

1. From same position as the gastroc stretch (p 56), with rear leg closer to front leg, (not as far back).

2. Bend the back knee, slowly sit lower into that leg, and feel the stretch deeper into the lower calf and Achilles tendon. Keep the weight on the front leg.

Test Your Flexibility

Stand facing the wall, in a staggered position. Place the toe of the front foot 2-4 inches from the wall. Bend the knee forward over the toes to touch the wall. Move the toes further away from wall and repeat.

As you become more flexible, you will place your foot further from the wall. The total distance is not as important as the symmetry between right and left sides.

If you are struggling at less than 2 inches, then this stretch is a priority. Your knee should be able to go over the front of the foot by 2-4 inches.

## 38. Front of the Lower Leg (Tibialis Anterior/ Peroneus Tertius, Toe Extensors)

This is a common area of irritation with shin splints on the front of the lower leg. This is also an important and often overlooked part of complete ankle mobility, especially after an ankle sprain.

Equipment needed: Exercise mat

1. Kneel on the floor with feet pointed out to the back. Lean forward to put your hands on the floor. This will allow you to control how much pressure you are applying to the front of the ankle.

2. Instead of trying to point the toes back further, lift the front of each knee a bit and you will feel the knee move forward slightly, bringing the front of the ankle closer to the floor. The general idea is to have the lower leg and the top of the foot flat on the floor.

Minimize compression by applying more pressure on the hands. When you are able to get the ankle flat, try to lift the knee off the floor more while sitting on the heel to put more tension on the muscle and the front of the ankle joint.

LOWER BODY

3. Putting the toes together and the heels out will stretch the peroneus tertius and the toe extensors.

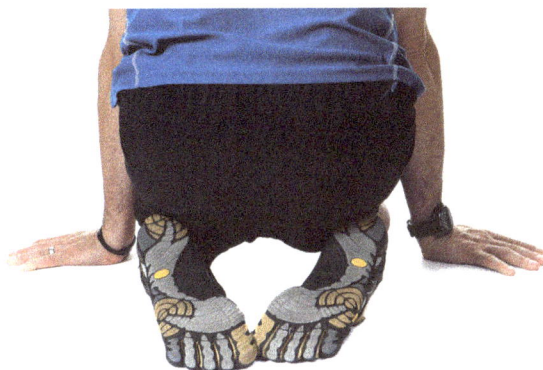

4. Putting the toes out and the heels in will stretch the tibialis anterior, which can be much tighter than the toes-pointed-in version.

5. Doing this stretch with the feet together, you will affect the peroneus tertius, the tibialis anterior, and the toe extensors.

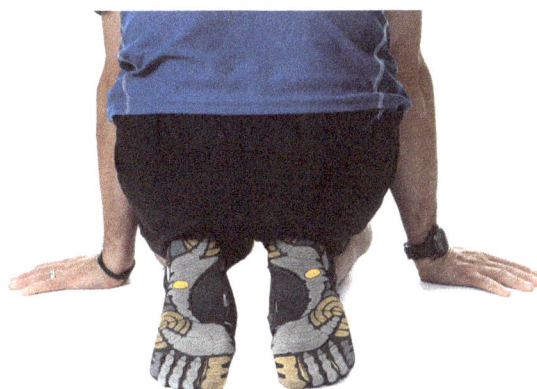

Hold for 30-60 seconds in each direction, 2X each side, or repeat 10X as a dynamic mobilizer. This can be very difficult after a sprained ankle.

# Feet and Toes

Equipment needed: Chair or bench

Footwear has a major influence on the toe joints so it's important to have flexible shoes that allow the toes to move freely.

Equipment needed: Bench or table and a chair

# Toe Flexors/Great Toe Stretch

### 39. Foot Intrinsic Muscles

These are the muscles within the foot that curl the toes under. To stretch these muscles, lift the heel of the back foot to bend the toes to 90 degrees, if you are able. Limit how much pressure you're putting on the toes using the other leg and a hand on a bench or table. Once the foot is straight up and down, put more pressure on the toes to the point that your entire body weight can be placed on the foot.

Hold for 30-60 seconds, 2X each side, or lift the heel up and down 10X as a dynamic mobilizer.

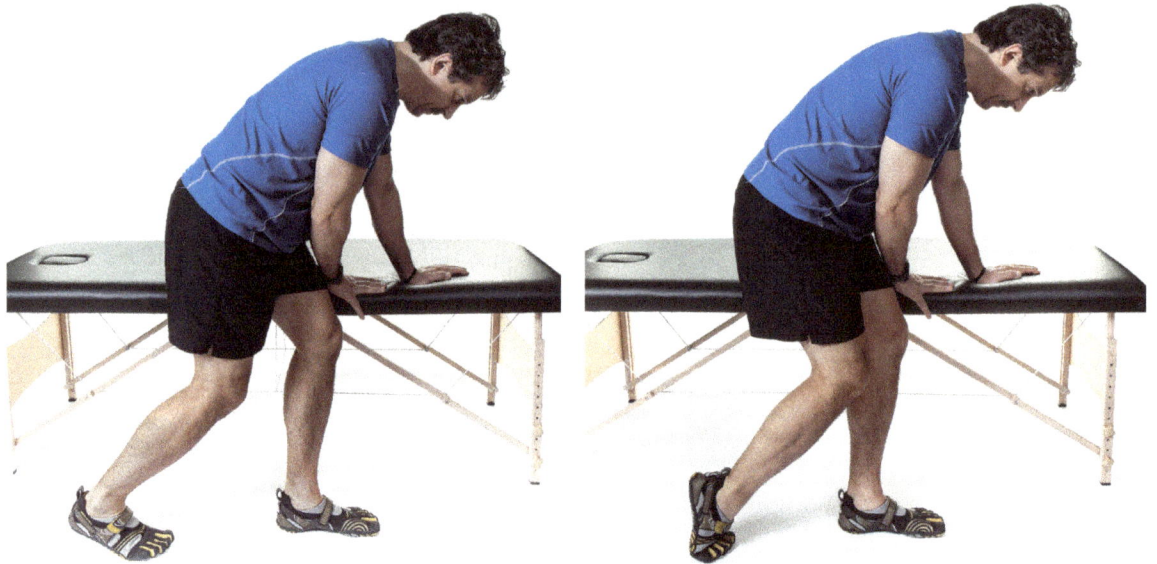

## 40. Foot Extrinsic Muscles

These are the muscles that control flexion of the toes—these are from the lower leg and cross the ankle, so the ankle must be involved in this stretch.

1. Place the foot on a chair with the heel elevated slightly.

2. Push the knee straight forward to bend the toes to 90 degrees, or as you are able. Do not drop the knee at this point. The toes are now bent to their limit and you can now sit the buttocks down on the heel if the knee and toes allow it.

3. Drop the knee toward the floor until you feel tension anywhere between the calf and the toes. Do not push to the point of pain—take your time and be patient.

Hold for 30-60 seconds, 2X each side, or lift the knee up and down 10X as a dynamic mobilizer.

# Toe Extensors

## 41. Foot Intrinsic Muscles

These are the muscles that pull the toes up. They are within the foot. From a seated position, take the ankle of one leg and place it just above the knee of the other leg. Keep the foot pulled up toward the shin slightly. With the foot up, bend the toes under with your fingers and hook the thumb under the ball of the foot for leverage. Use either hand.

Hold for 30-60 seconds 2X, or repeat 10X as a dynamic mobilizer.

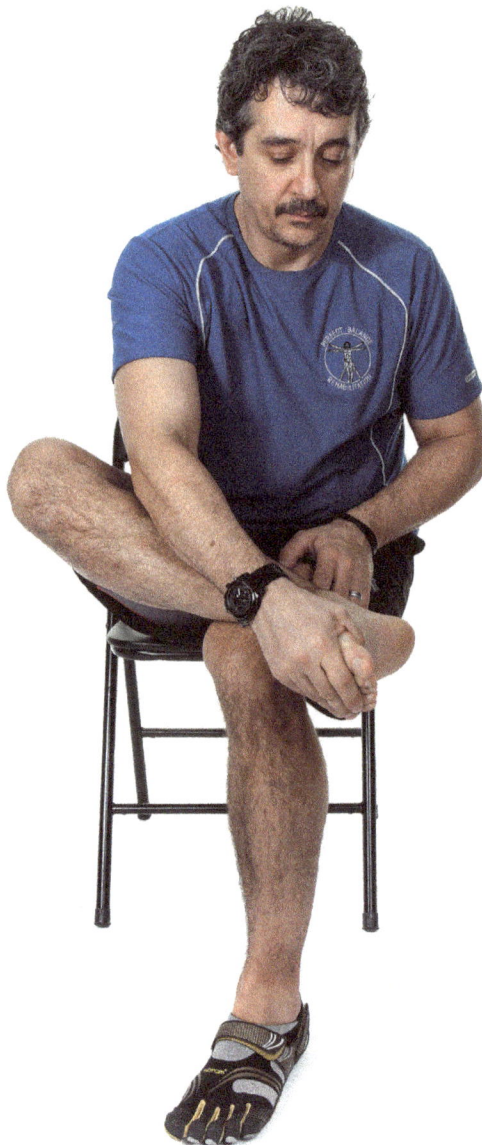

**LOWER BODY**

## 42. Foot Extrinsic Muscles

These muscles lift the toes, and they come from above the ankle. From a seated position, take the ankle of one leg and place it just above the knee of the other leg. Keep the foot pointed down and apply the stretch with your hand.

Hold for 30-60 seconds 2X, or repeat 10X as a dynamic mobilizer.

# Upper Body

## Chest Muscles (Pectorals)

The pectorals (pecs) are chronically tight and short in most of us as our lifestyle encourages a rolled forward shoulder posture, such as when we are at a desk or driving.

Equipment needed: Wall or leg of a squat rack or an inner corner of the wall

### 43. Single Arm Pec Stretch

1. Stand with forearm and hand against the wall, outside leg forward and inside leg slightly staggered back with knees bent and holding a pelvic tilt. Put the arm out to the side with elbow bent to 90 degrees and hand pointing up to ceiling.

2. Push the inside shoulder toward the front and turn the outside shoulder away, toward the back. Make sure your elbow is slightly below the shoulder and shoulder blade is pulled down.

Try not to poke head or pelvis forward keeping the spine in neutral position. Place the elbow above, below, and at the level of the shoulder. Hold for 30-60 seconds, 2X each side, or repeat 10x as a dynamic mobilizer.

## 44. Bilateral Pec Stretch - Middle

Stand in the corner of two walls, and press the chest into the corner. Maintain neutral spine with the chin back and holding a pelvic tilt. Keep your forearms vertical with the elbows slightly lower than shoulder height. Keep the feet under the shoulders and lean slightly to get light pec pressure. Having the feet too far back will engage the pecs, creating a contraction rather than a stretch. Follow the breathing pattern shown below.

Breathing to elicit a stretch:

1. Move forward until you feel mild tension in the chest, and hold the position.

2. Take 3 slow deep breaths into the chest. As you breathe in, the chest muscles will feel tighter, and as you breathe out, they will release.

3. On third breath out, lean closer to the corner. Hold again.

4. Repeat 3 deep breaths and go forward on the third breath out.

5. Take arms down and relax.

6. Choose another shoulder angle and repeat 2X at each position.

## 45. Lower Pecs

Repeat previous stretch with elbows placed 20-30 degrees above the level of the shoulder.

## 46. Upper Pecs

Repeat previous stretch with elbows placed 20-30 degrees below the level of the shoulder.

## 47. Middle Back (Rhomboids/Middle Trapezius), Back of The Shoulder (Posterior Deltoid)

The middle back muscles contribute significantly to posture, are used for rowing and pulling things toward you from the front, and are closely associated with the biceps muscles.

Equipment needed: Wall or leg of a squat rack

*Beginner*

1. Place the fingertips of one hand on the shoulder of the same side. Lift the elbow so it is straight in front of the shoulder at 90 degrees of flexion or slightly below.

2. Pull the elbow across the chest with the other hand, level with the opposite shoulder, while turning the head in the same direction to prevent the forearm from putting pressure on the throat. Pull until you feel tension in the back of the shoulder. Pressure in the front of the shoulder may indicate pectoral muscle tension and may need to be released first. You should feel the stretch across the back of the shoulder or into the middle back near the shoulder blade close to the spine. Try not to arch the back.

3. As your range of motion increases, pull the elbow across the chest by placing the wrist on the outside of the elbow and flex the elbow on that side to increase the stretch.

The reason for the hand on the same shoulder is to encourage better shoulder mechanics and allows a more direct stretch to the targeted area without your shoulder coming up to your ear.

*Advanced*

1. Place the elbow and shoulder of the arm being stretched on a wall, or just the elbow on the leg of squat rack, with shoulders perpendicular to the wall and the elbow in the same plane as the shoulder. Place the feet relatively close to the wall, shoulder-width apart.

2. Turn the body and feet to try to touch the opposite shoulder to the elbow. The foot on the opposite side of the shoulder being stretched should be slightly back so that you can use the leg to push the shoulder into the elbow. The foot on the same side as the stretched shoulder and the shoulder and forearm on the wall provide support. The other hand can be placed across the stomach. Try to maintain a neutral spine. Lean into the wall only as much as you need to affect a stretch. Once you can touch the shoulder to the elbow, it is now simply a matter of maintenance.

Hold for 30-60 seconds, 2X each side, or repeat 10x as a dynamic mobilizer

## 48. Middle Back (Latissimus Dorsi (Lats)/Lower Trapezius)

These are the muscles used to do a chin up and they are also used to slow down the arms at the end of a swing of a bat, racquet, or golf club. They also slow the arm down as you release from throwing something. These are the muscles that allow us to row a boat, paddle board, canoe, or kayak. This is one of our favorite stretches and feels great.

Equipment needed: Edge of door frame, wall, or leg of a squat rack

With our tendency to sit stooped, with the shoulders rounded forward and down, the latissimus dorsi muscle that connects the upper arm to the pelvis also becomes very tight, along with the lower trapezius muscle. This will have a huge influence on the mechanics of the shoulder joint.

1. Place a hand on the wall straight in front of the same side shoulder with a slight bend at the elbow. All other foot and hand placements use this hand as a reference point. Place the same side foot near the wall directly below the hand. Reach the opposite hand directly above the first-hand placement with this hand high enough to make you reach a bit. Reach the other foot behind the first foot placement with a small amount of weight on the toe and the heel slightly elevated off the floor. The hands and feet will all be in a straight line.

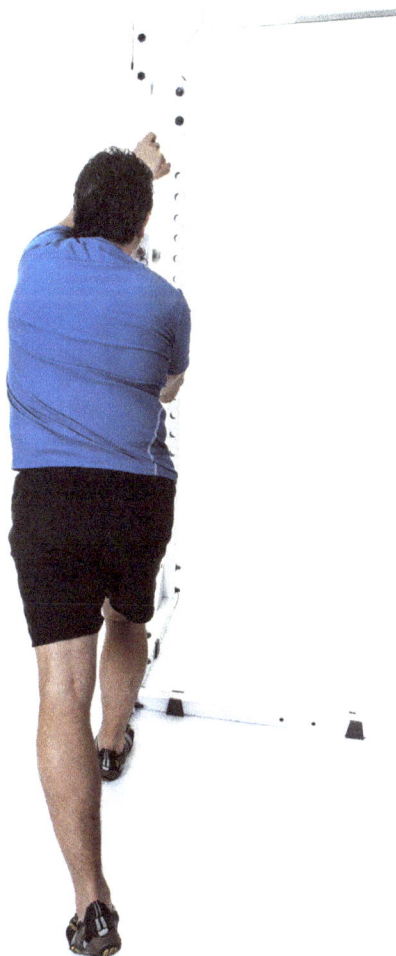

2. Push the shoulders and hips to the side and away from the lower hand and front foot and push the heel down to intensify the stretch. If the back heel makes contact with the floor, reach the top hand up a bit higher while lifting the back heel again, and repeat the process.

   A door frame, pole or a line on the floor/wall may be used as another point of reference. The lower hand and front foot control the body weight and only enough pressure on the top hand and back foot to feel the stretch. The entire body is in the shape of a big "C" from the hand to the foot. Switch and stretch the other side by reversing the hand and foot positions. Use only enough pressure on the top hand and back foot to make it a comfortable stretch.

Hold for 30-60 seconds, 2X each side.

Note: Because of the relationship of this muscle to the ribs, taking 4-6 deep breaths and pushing further with each exhale makes this stretch much more effective. As always, you can also repeat this stretch 10x as a dynamic mobilizer.

# Shoulders

The shoulders are extremely mobile, so it is vital to maintain flexibility and strength throughout your lifetime. Shoulder injuries in overhead sports are very common and tricky to heal and rehabilitate.

Equipment needed: Cord, wall, or leg of a squat rack

## 49. Internal Rotators (Subscapularis and Anterior Deltoid)

1. In standing position, reach overhead with an arm and drop a skipping rope, belt, or cord down the back while bending the elbow to reach the hand down toward the middle back to a comfortable position.

2. Reach behind with the other hand and grab the rope near your waist. With the upper arm stable and the elbow up near the ear, pull down with lower hand to pull the upper hand down the middle of the back until you feel tension. You may feel tension in the upper triceps region and into the armpit.

Hold for 30-60 seconds, 2X each side, or repeat 10X as a dynamic mobilizer.

Another effective external rotation stretch is to keep the elbow into your side and bend it to 90 degrees. Rotate the upper arm to push the lower arm out to the side. Grab a door frame, pole, leg of a squat rack, or even the wall with the hand, and turn the body away from the hand. Push only until you feel a light tension. The goal is to get your hand straight out to the side away from the body. Keep the core slightly engaged to prevent arching the lower back.

Hold for 30-60 seconds, 2X each side, and take 4-6 deep breaths, pushing further with each exhale, or repeat 10X as a dynamic mobilizer.

## 50. External Rotators (Supraspinatus/Infraspinatus)

*Beginner*

1. In a standing position, reach overhead with an arm and drop a skipping rope, belt, or cord down the back while bending the elbow to reach the hand down toward the middle back to a comfortable position.

2. Ensure that the elbow of the lower hand is pulled tight into the body and the hand is on the opposite side of the body. Many people will try to pull the hand up the middle of the spine but this will only jam the shoulder.

3. With the upper hand, pull the rope up to lift the lower hand in an arc of motion up the outside of the body until a comfortable stretch is felt in the front of the shoulder. When full range of motion is attained, the hand will be in the middle of the spine or as close as possible.

Hold for 30-60 seconds, 2X each side, or 10x as a dynamic mobilizer.

*Alternate method:*

Grab the leg of a squat rack or outside edge of an open door.

Squat down to allow the hand to creep up the back until you feel tension in the shoulder. Release the grip and then stand up, taking the hand up with you. Grip again, squat again, and repeat 2-3X within 30-60 seconds, 2X each side.

## 51. External/Internal Rotation Stretch

*Advanced*

Once you can reach down the back from above with one hand, and reach up the back from below with the other hand and touch the fingers together, you don't need to stretch any further—just maintain this level. You may never get this far and this may be due to your unique anatomy. If you have pain and it won't go away, please seek professional assistance.

## 52. Flip-Flop Mobilizer

This stretch is a unique way to increase internal rotation in an active fashion and it can be used with rotator cuff issues.

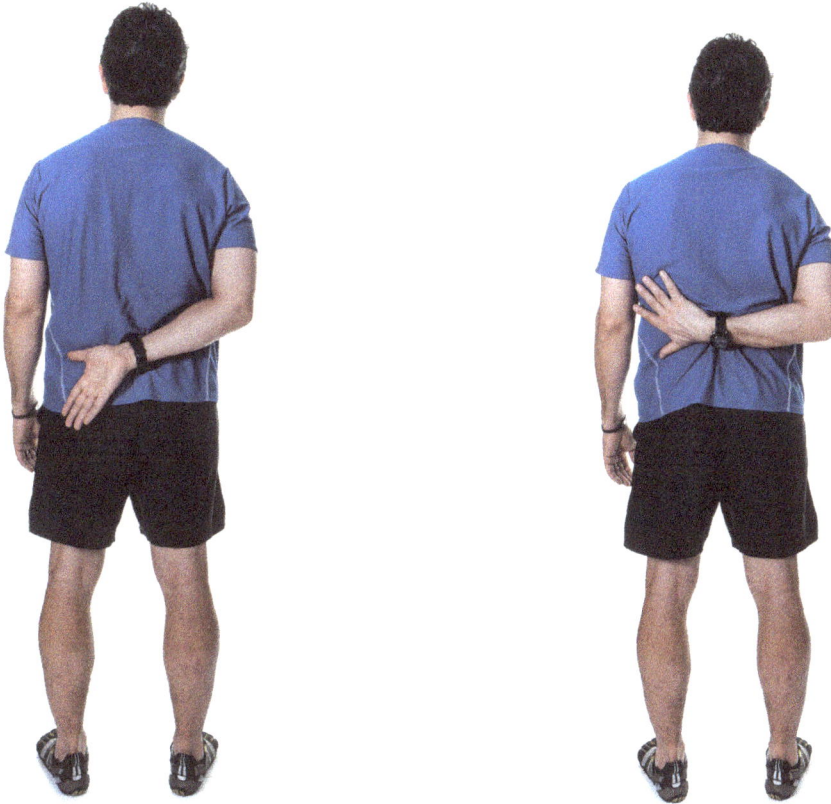

1. Reach across the back with the hand. You may only get as high as the buttocks or even the tailbone, but push only until you feel very light tension.

Although the pictures show an open hand, we would prefer that you make a fist and slowly roll the hand and wrist over the thumb, and try to touch the little finger on the back, keeping the hand in a fist. This is the flip. Lower the fist if this is painful, until the little finger can touch with a tug but with no pain.

2. Let the wrist roll down again to release the tension. This is the flop.

3. Repeat several times until the tension goes away (10-20 times). If tension doesn't go away, try later after some ice or heat on the shoulder. The type of injury will dictate how far you can push. If the tension goes away, flop the hand and raise it up a tiny bit more and work your way up the back. In the flip position, don't expect to touch the little finger higher than the bottom of the opposite shoulder blade. In the flop position you may strive to touch the middle of the spine just below the neck.

## 53. Back of Upper Arm (Triceps)

These are the muscles that, in conjunction with the chest muscles, are used to do push ups or a triceps press-down. Involving both the elbow and shoulder will create a much more effective stretch for the belly of the triceps muscle.

Equipment needed: Wall or leg of a squat rack

1. Reach up a wall or leg of a rack with a bent elbow and place the fingertips on the back of the same side shoulder.

2. Reach in front of the neck to grab the wrist of the arm being stretched and pull it down into the shoulder. Keep the elbow of this arm down near the chest. Lean the head over to the side to assist holding the hand down on the shoulder. Reach the elbow as high as possible or squat just a bit to increase the intensity of the stretch. Placing the back of the upper arm and armpit on the wall with the hand flat on the back of the shoulder is the end goal. Use caution to avoid irritating the shoulder joint.

Hold for 30-60 seconds, 2X each side, or repeat 10x as a dynamic mobilizer.

## 54. Elbow Mobilizer

Equipment needed: Wall or leg of a squat rack

This is a very joint-specific stretch. It is used when it is difficult to bend the elbow to touch the finger tips to the shoulder with the elbow down by the side.

1. Bend the elbow with upper arm by the side so the fingertips are as close to the front of the shoulder as possible. Place the lower arm and wrist on a wall.

2. Press the shoulder into the hand until tension in the elbow joint. This is more of a joint issue than a muscular one, and you will likely feel something in the joint. Ideally, the shoulder will be able to touch the hand. Stand with feet shoulder-width apart and place the foot on the same side back a bit which can be used to push the shoulder forward if the elbow is pain free.

Hold for 30-60 seconds, 2X each side, or repeat 10 x as a dynamic mobilizer.

## 55. Front of Upper Arm (Biceps)

A very common throwing or serving (racquet sports or volleyball) injury to the front of the shoulder and upper arm affects the long head of the biceps. As with the triceps, this muscle crosses two joints, so to be most effective, we need to incorporate the shoulder and the elbow.

Equipment needed: Wall, table, desk, or bench. The ideal is an incline bench with a weighted bar.

1. Sit on a ball or chair and reach back to grab a bar or place your wrists on the top of a table with the palms up. Ensure that you are balanced and secure on the ball so that you don't have to use the arms to balance.

2. Squeeze the shoulders back and keep the arms straight. Leaning forward and rolling the ball back will reduce the pressure.

3. Roll the ball forward until you feel a comfortable tension in the muscle belly of the biceps. If you feel the stretch at either end of the tendons at the top or bottom, the tendons are compromised, so back off a bit—you will feel the pull in the weakest link.

Push until you feel mild tension anywhere, from the front or top of the shoulder to just below the elbow, and all points between.

One of the best places to do this stretch is on an inclined bench as it is much easier to control the

stretch and is extremely stable. Grab the bar and slide the buttocks up and down the bench to control the stretch with the legs.

This stretch can also be done by placing the hands up high on a wall. Squeeze the shoulders back and then squat. You could also try a backstop or a fence. Try to do the stretch with the palms down as well; this will affect a different part of the shoulder joint.

Hold for 30-60 seconds, 2X, or repeat 10X as a dynamic mobilizer.

# Forearm

These are the muscles that control the wrist bending up and down and side to side, turning the hand palm down and palm up as well as bending and straightening the fingers.

## 56. Tennis Elbow Stretch

This is a stretch of the muscles that extend the fingers and hand (lift the hand and fingers with palm down) that are commonly used and often injured hitting a backhand in racquet sports from which the name is derived. When this area is injured, you will feel pain on the bump on the outside of the elbow and into the forearm or wrist.

Equipment needed: Chair or bench

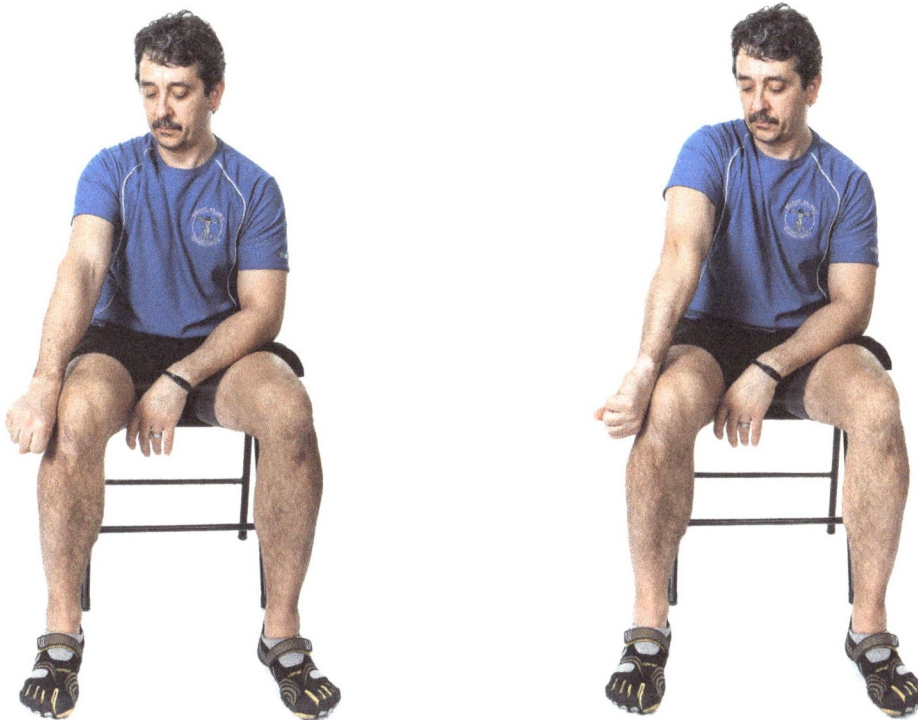

1. From a seated position, place the hand to the outside of the thigh on the same side and straighten the arm.

2. Turn the hand to point the thumb toward the floor.

3. Put the hand in a fist and flex the wrist (bend to the outside) until slight tension is felt on the top of the hand, wrist, or forearm.

4. Reach the other hand over the flexed wrist and grab the hand so two fingers are on one side of the index (pointer) finger knuckle and two fingers are on the other side of the knuckle. Hold the thumb down on the stretched side with the thumb of the hand controlling the stretch. The fingertips of the index and middle finger should be wrapped around the edge of the index finger of the hand being stretched. The knuckles of the target hand will be in the palm of the hand controlling the stretch.

5. Hold the target hand secure and bend the elbow of the target arm out to the side to increase the stretch as this will provide better leverage and control than pulling the hand back.

6. Return the elbow back to the arm straight position, flex the wrist more and turn the hand up to the ceiling with the control hand until you feel tension in the forearm, then bend the elbow again. Repeat 3 times. Compare sides and try to keep them as even as possible.

Hold for 30-60 seconds, 2X each side, or repeat 10X as a dynamic mobilizer.

# Wrist

## 58. Wrist Flexor Stretch

This is purely a wrist stretch and does not involve the fingers to any great extent, but they may still be affected if the hands are very tight. These are the muscles that lift the hand with the palm up.

Hold for 30-60 seconds in each position in the list below. Lean forward to intensify the stretch and lean back to decrease the stretch. Palms are down and arms are straight.

Equipment needed: Bench or table, or kneel on the floor on an exercise mat

1. Place thumbs together and fingers out to form a W.

2. Place fingers straight and thumb tips touching to form an upside- down M.

3. Touch the thumb tips and index fingertips to form a triangle.

## 57. Golfer's Elbow Stretch

This stretch is for the muscles that flex the wrist and fingers (lift the hand with the palm up and make a fist) which can be injured in golf by hitting the ground too hard or hitting a tree root, causing the hands to go ahead of the club face and forcing the wrist into extension of the lower hand on the club. When this area is injured, you will feel pain on the bump on the inside of the elbow and down into the forearm.

Equipment needed: Chair or bench

1. From a seated position, open the hand and place the forearm on the inside/top of the thigh, palm up with the elbow bent.

2. Place the palm of the other hand on the fingers of the stretched hand with the index and middle fingers on the thumb.

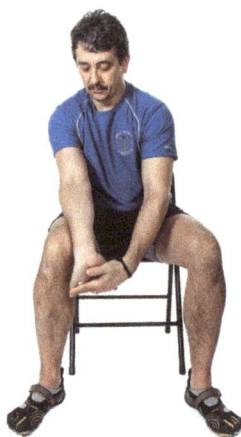

3. Straighten the arm of the stretch side while gently pulling the thumb and fingers back until you feel light tension.

4. Pull and turn the stretch side hand toward the little finger to increase tension a bit more.

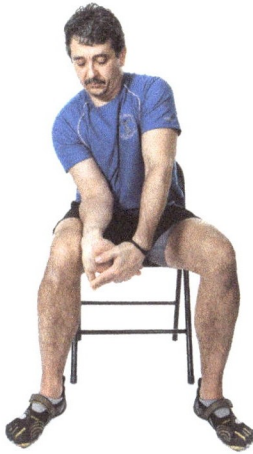

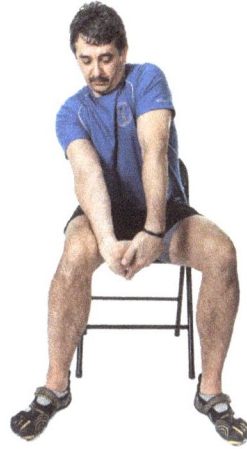

5. From this position, bend the elbow down to control the intensity of the stretch while holding the hand secure.

6. When you feel a comfortable stretch, hold for 30-60 seconds. Straighten the elbow to release the tension, pull the hand back and inward again to a point of tension, then bend the elbow again, and repeat 3 times. After the third time, the wrist has probably reached maximum range of motion for that session. You may feel tension anywhere from the fingers up to the elbow.

If you feel a sting at the wrist, this could be a partially entrapped nerve, possibly indicating the start of carpal tunnel syndrome. Seek professional assistance.

Another good flexor stretch: Place one hand flat on a bench with the other hand forward on the bench to provide support. This stretch involves both the wrist and the fingers, which makes it far more effective. Since the advent of grasping tools, sports equipment, and even typing, this is an extremely important stretch.

1. Bend the wrist to 90 degrees, if you are able.

2. From this position, lift the palm while keeping the arm straight and then lean forward toward the support hand while lowering the shoulder to the bench. A great overall stretch of the forearm flexors that allows great control.

## 59. Wrist Extensor Stretch

These stretches are a bit tricky and you may have to lean forward over the wrists on a bench or start on the floor to allow better control if the wrists are particularly tight. Place the hands palms up, fingers point toward you, bend the elbows a bit to release tension and then slowly try to straighten the elbows to increase tension before attempting to lean back. Limit the pressure as necessary and apply more pressure on the wrist joint as able.

1. Place fingers in to form a triangle, with index fingertips and thumb tips touching.

2. Keeping thumbs in contact rotate wrists and hands to pull the index fingers away from each other until fingers point straight toward the legs to form the letter M.

3. Place fingers out and thumbs together to form an upside-down W.

Hold each position for 30-60 seconds. This is also mostly pure wrist, but if you want to involve the fingers, make a fist first, but be very careful as this can get tight quickly.

# Appendix 1: Steps in Healing

The terms collagen, viscoelasticity, elongation, hysteresis, creep, and load relaxation will be explored in this section. If you want to get a bit technical and become more aware of the process of healing, this appendix is for you.

Connective tissue in and around muscles is non-contractile and the main substance in these tissues is known as collagen. Collagen has viscoelastic properties in that it can change shape and be deformed, but viscous tissue will deform and stay that way (like playdough). Fortunately, normal tissue also has the properties of elasticity, which means it will return back to its original length (like an elastic band). When deformation or lengthening of this tissue is sustained, recovery is slow when the force is removed.

When doing a stretch, static or dynamic, we are trying to elongate the tissue. Stretching involves lengthening, and a contraction involves holding or shortening. (There are also eccentric contractions, but that is another conversation.) When doing a static stretch, the muscles are down-regulated and need to be as relaxed as possible. "Creep" describes the ability of tissue to elongate over time when a constant load is applied to it. "Load relaxation" means that less force is needed to maintain this new length. "Hysteresis" is the amount of lengthening a tissue will maintain after a cycle of stretching (deformation) and then relaxation. This effect is reported to last about four hours, so is static stretching really about the muscle or is it more about fascia, ligaments, joint capsules, vascular and neural vessels, and skin tightness resulting in resistance to normal fluid motion? We are saying that it is the latter.

After a static stretch, all of these structures, including the muscles, have changed their length, and consequently, the muscles and joints now have to perform a slightly different function. It will therefore feel a bit unusual to move after a static stretch due to more fluid motion and neural inhibition.

The primary consideration with respect to static stretching involves injury. All tissues, including muscle, tendon, ligament, and including on or below the skin, go through the same healing process after injury. There are four overlapping stages involved in tissue healing.

## Stages of Healing

1. With injury, disruption of the cells will result in bleeding and the release of chemical mediators from the disrupted cells that, in turn, signal the body to secrete inflammatory elements into the area. Stage I is HOMEOSTASIS, in which vasoconstriction happens to stop the bleeding. Fibroblasts, which make

collagen as well as fibrin, are released into the area to form a clot and provide early support for the injured area, also known as a scar. Coagulation and homeostasis are in the first stage of healing.

2. White blood cells, some called macrophages, are also released to clean up the injured area by a process called phagocytosis. The ACUTE INFLAMMATORY STAGE, stage II, can result in significant inflammation. The presence of macrophages also results in the attraction of fibroblasts to produce more collagen. This stage lasts about two weeks.

3. Stage III is TISSUE PROLIFERATION AND REPAIR, where production of collagen produces fibrotic adhesions to bind torn tissue in more serious injuries. This starts at about day three and continues for four to six weeks. During this stage, the collagen starts to contract to further stabilize the area but it still has very weak hydrogen bonds, is poorly organized, and poorly vascularized. Gentle stress and strain, like stretching, help to stimulate the production of more organized collagen and progression to the final stage.

4. Stage IV is TISSUE REMODELING, in which movement stimulates synthesis, lysis (absorption), and reorganization of the scar. Stress causes a piezoelectric effect which helps the collagen to realign itself and form stronger covalent bonds rather than weak hydrogen bonds. During this stage, normal activities can be resumed early and gradually, but healing is affected by the amount of scarring and blood flow. This stage can last ten to twelve months, but it can last longer.

Formation of adhesions is a primary purpose of tissue healing, and a primary function of stretching is to reduce adhesions. Unfortunately, adhesions can bind the muscle to itself and to other surrounding tissue. This can cause restricted range of motion, joint instability, pain and potential lifelong disability from ligament-deficient joints, even with adhesion formation present. Nerve innervation, circulation, and proper biomechanical and nutritional exchange can also be affected by adhesion formation. After an injury, the muscles will contract (spasm) to stop movement of a joint or the muscle itself to minimize pain. If held long enough, contractures can result which, if not "stretched out," will inhibit movement.

Static stretching should be slow and controlled. If done quickly, a stretch reflex via the muscle spindle is activated, triggering a reflexive contraction of the muscle being stretched, which is undesirable. However, if the muscle is held under low tension for longer than six seconds, another receptor called a golgi tendon organ is stimulated, which relaxes the muscle as well as having an inhibitory effect on the muscle spindle.

Static stretching will help to prevent adhesions from producing disability and should be started gradually in stage II, when things are still nice and soft, and more amenable to change. If you don't move it, you will lose it. This explains the early mobilization of injuries and the surgical interventions used today, as opposed to the "don't move injuries," lengthy hospital stays, and immobilization that we embraced in the not- too distant past.

According to the late renowned body worker Ida Rolf:

- Habits become structure.
- Structure becomes posture.
- Posture creates the fault lines of the body.
- Posture causes or contributes to pain (symptoms).
- By changing posture, pain and symptoms may be resolved or reduced.

# Sources

1. Smith, Clive. (1992). "How tissues heal." Insurance Corporation of British Columbia.

2. Hamilton, Andrew. (2019). "Flexibility and stretching: The importance of keeping flexible. Strength, conditioning and flexibility." Peak Performance.

3. Martin, Molly. (2017). "Golgi Tendon Organs and Muscle Spindles Explained." American Council on Exercise.

# Appendix 2: Application of the Stretches to Daily Life

The following is a selection of stretches that will help alleviate some of the symptoms of a variety of common issues that we have observed and that people have shared with us over the years.

1. Rounded shoulders/head-forward posture/headaches

    a. Chest (p 66-68)
    b. Neck rotation/side flexion (p 34-36)
    c. Retractions (p 30-31)
    d. Neck and middle back mobilizers (Cervical and Thoracic spine extension) (p 27)
    e. Extension mobilizer (p 26)

2. Flat feet/plantar fasciitis/sore arches

    a. Toe Flexors/ Great toe stretch (p 62,63)
    b. Lower Leg/ Calf (Gastroc and Soleus) stretches (p 56 and 57)
    c. Tibialis Posterior stretch (p 58)
    d. Front of Lower leg (p 60,61)
    e. Peroneal (p 57)
    f. Toe extensors (p 64,65)

3. Low back pain/ tightness

    a. Pelvic tilt (p 15,16)
    b. Front of hip (Hip Flexors) (p 38-41)
    c. Outside of hip (TFL) (p 52,53)
    d. Buttocks (Gluteals) (p 42- 45)
    e. Front of thigh (Quadriceps) (p 46,47)
    f. Back of thigh (Hamstrings) (p 48-50)
    g. Sciatic nerve tensioner (p 51)

4. Forearm Injuries

    a. Golfer's elbow (p 86)

    b. Tennis elbow (p 83-85)

    c. Wrist flexors (p 88,89)

    d. Wrist extensors (p 90)

5. Shoulder injuries

    a. Middle back (Lats) (p 71,72)

    b. Chest (pectorals) (p 66-68)

    c. Middle back (Rhomboids) and back of shoulder (posterior Deltoid) (p 69,70)

    d. Front of upper arm (Biceps) (p 81,82)

    e. Triceps (back of upper arm) (p 79)

    f. Shoulder internal rotators (p 73)

    g. Shoulder external rotators (p 74-76)

    h. Flip flop (p 77)

6. Just 'cause it feels good.

    a. Chest stretch (p 66-68)

    b. Front of hip (Hip Flexors) (p 38-41)

    c. Buttocks (Gluteals) (p 43-45)

    d. Prone extension (p 23,24)

    e. Child's pose (p 18)

    f. Extension mobilizer on the ball (Cervical, Thoracic, and Lumbar Spine extension) (p 26-28)

    g. Lying torso rotation (p 13,14)

    h. Wrist flexor (single arm version on bench) (p 89)